建筑立场系列丛书 No.52

地域文脉与大学建筑
University Buildings in Context

C3

汉英对照

(韩语版第368期)

韩国C3出版公社 | 编

史虹涛 时真妹 马莉 张琳娜 周一 王忠倩 | 译

大连理工大学出版社

地域文脉与大学建筑

增建与拆除

C3 建筑立场系列丛书 No. 52

University Buildings in Context

Addition and Subtraction

地域文脉与大学建筑
University Buildings

无论大学与历史遗留建筑、标志性建筑、一个地区或一个国家的联系有多紧密，这个词的概念都是以不断的发展、变迁、扩大或是缩小为基础的，且它的本质就是志同道合的人组成的共同体。大学以其价值观体系和求知探索为第一要务，吸引着那些愿意分担此责任的人们。从这个意义来讲，大学超越了时空的限制。它们还充当社会压缩器，变多元化为共同体。

大学活动必定需要场所，而建造这样的场所很有可能需要超越地理的维度。然而，大学可以让世界各地慕名而来的人们通过实地或互联网的方式访问这里。大学的校址与首都一样具有价值的源动力，可以产生广泛的效益。校区、城市、甚至国家都能因此产生神奇的魅力，它不仅仅用于文化创造，还能进一步对地区经济产生实在的影响。

建筑兼顾于功能与空间，在满足实用性的同时，首要考虑的是空间品质。建筑要符合场地的需求，这需要与所属社区协调并达到一致，以此来决定与现存自然景观或城市肌理的融合度和不同特点。最后，建筑还要传达大学的目标和价值观，能够充当教学工具，用自己形式和功能性语言，传递观点，激发想象，促进创新。

本文个案研究中选取以下案例，目的是展示区域或地方背景下大学建筑的理念。

However tied universities might be with a heritage building, an iconic establishment, a place or a nation, the very word is actually founded on the notion of an ever-evolving, shifting and moving, expanding and contracting, association of individuals who are incorporated into a body of shared aspirations and function. The university is, first and foremost, a set of values and an endeavor of the exploration of knowledge, appealing to those who are apt to take on the shared burden of the task. By this definition universities transcend the limitations of place and space. They also act as a social condenser, turning diversity into community.

By necessity though, this activity is bound to happen somewhere. It is also likely to invest this "somewhere" with appeal that extends further than mere geography. Likewise, the crowd that is attracted to that "somewhere" now comes from all over the world, in physical or digital space. Seen as a generator of such interest, the locality of the university amounts to a capital that establishes value on a wider scale. Campuses, cities, even nations, achieve even mythical appeal because of this activity, which is not only limited to cultural production but also extending to an actual economic impact on the place.

Architecture comes in at the gap between function and place, reinstating spatial qualities as the foremost property of an otherwise utilitarian endeavor. It responds to the requirements of the site, it negotiates the terms of accessibility of and to the attached community, and it decides the level of integration or distinction to the existing natural landscape or urban fabric. Finally it communicates the university's aims and values, or it can act as a pedagogical tool by using its own formal and functional language to convey ideas, ignite the imagination, or foster creativity.

The selection of the case studies for this article was made in order to showcase the idea of building for the university, in the context of a regional or local condition.

in Context

地域文脉：大学和地区

根据定义，大学代表一个集体。大学一词源于拉丁文"Universitas"，指团体、全体、机构，或是有共同追求的教师和学者的组织（Universitas Magistrorum and Scolarium[1]）。有趣的是，这个词以其学术活动概括了团体的精神，但并没有考虑希腊语"akadēmeia"所指代的机构空间的概念[2]。从这个意义上讲，大学是以吸引志同道合的人为功能的中心。

大学以其管理自主和学术自由[3]的基本特征，徘徊于现实和宗教之间。也因此让大学在数百年来广纳贤客，成为挑战知识局限，探寻研究之道的中心。此外，大学还摆脱了场地和空间的限制，以致一些最有趣的大学都隐藏在不起眼的地方，老师和学生应学术活动的要求而自由往来。总之，大学是一个协会。尽管大学的本质是不断发展，但它仍然需要特定的场所[4]。场所要考虑周围环境，正是基于这点，成就了大学对当地的意义和影响力。

大学的制度化涉及选址的问题。位于巴黎的"老师和学者的大学（协会）"是最古老的协会机构，在建立之初曾面临搬迁的威胁。1229年的罢工发生在协会从巴黎搬走[5]之后，明确地展示了学术工作被剥夺，以至于老师和学生考虑另谋出路，城市是怎样失去吸引力的。经济活动的缺失，将导致国家和附近区域（本案例中指拉丁角）受到巨大的经济压力。随着学生从欧洲各地汇集此地，大学又变得重要起来。

当下的情况又是如何呢？现实是大学正在进行"前所未有的扩张"[6]，先进的通讯技术信息的广泛传播，以及学生和老师的流动都能超越场所的限制。学生不用亲自到校就可远程学习知识，补充、完成大学课程。学者可以进行国际合作，通过期刊、会议和教学交流，在全球范围内发表他们的研究成果。院系合作通过联合项目和联合会得以实现。发展中国家和曾经的殖民国家可从世界各地招聘教师，招收学生，而不必

In the Context of Place: Universities and Regional Localities

By its very definition, the university stands for a collective. Deriving from the Latin word "Universitas", the word refers rather to a whole, a total, an institution and a body of teachers and scholars ("Universitas Magistrorum and Scholarum"[1]) sharing a common endeavor. Interestingly enough, the word encapsulates the spirit of corporation through scholarly activity instead of a place of instruction, that the Greek akadēmeia would stand for[2]. In that sense, Universities act as centers of gravity for a function that, in turn, attracts a collection of individuals into a recognized body of shared aspirations.

Standing in the gap between secular and religious authority, the university is fundamentally characterized by its corporate autonomy and academic freedom[3]. This has allowed it to act through the centuries as a welcoming hub for challenging the limits of knowledge and the methodology of inquiry. Likewise, it has existed past the limitations of given place and space; indeed some of the most interesting universities were sheltered in unassuming environments while the mobility of teachers and students is normally thought in direct relationship to the appeal of a scholarly practice. All in all, the university is a guild. Yet, in spite of its ever-evolving nature, it is necessarily bound to a certain situatedness[4]. It needs to take place in context, and it is at that point where it becomes impactful and meaningful to a place.

The very institutionalization of the university is attached to a question of settlement. Indeed the founding act of the "University [cf. association] of Masters and Scholars" in Paris, argued to be the oldest institution of the sort, has taken place on a threat to leave. An account of the strike in 1229, followed by the migration of the guild from Paris[5], shows explicitly how the city stands to lose much of its attractiveness if the scholarly work was to seize and the masters and students assumed their practice elsewhere; likewise a large economic strain would apply to the state, as well as to the immediate neighborhood (in this case the Quartier Latin), due to the loss of economic activity. With students coming in from all parts of Europe, place suddenly becomes important.

How does this mirror the current situation? For one thing Universities are now subject to "an unprecedented expansion"[6]. Similarly, the wide dispersion of information through advances in com-

Drawing©Heatherwick Studio

再将学生派遣到西方国家。已有院校也不仅限于在本土"经营"和扩大影响。这只是全球社会乃至全球知识首府的局部呈现。尽管如此，我们并没有偏离交流活动的基本模式，比如，在中世纪大学的"使用拉丁语这门通用语言在教师、学生之间交流思想"的模式，和早期现代大学[7]的"恢复思想启蒙运动时期的传统知识和科学发现等领域的展开合作"的活动都证明了这点。

建筑又是什么情形呢？它该如何构建，拓展校区理念并实现自己的愿景呢？不要忘记，建筑最终是要服务于需求的，如果其建设只局限于人体工程学特性就不太可能使建筑满足需求的目的，前面我们已经讨论过，建筑的功能不仅仅是教学。按照我们描述的架构，对建筑最起码的要求是尽力而为：生成空间，聚集群体，创造意义。要将空间转化为居所，建筑需要因地制宜，而不是削足适履。据此，我们研究的项目需与当地情形相得益彰并建立可持续的关系，要成为一处景观，它要协调好与当地社区的关系，以及与历史和地理场地的关系。此外，建筑具有向世人宣传的作用，邀请那些心怀共同抱负的人来到大学。在此背景下，我们研究的项目要彰显院校的个性和价值观，用最明了的形象展现它的立场。最后，建筑最重要的功能之一是为人们和活动提供场所。我们所研究的建筑设计初衷是为教学、讲座、实验、服务等提供适合的空间，因此通过嵌入过渡空间、打破边界、创造未知的交锋来推动社区的发展。

由CRAB工作室设计的Abedian建筑学院位于澳大利亚昆士兰州，邦德大学西北角。Abedian建筑学院高三层，发展为一串彼此相连的荷叶结构。整个体量由一系列的"风斗"铰接而成：混凝土浇筑的凹形区有节奏地排列在中央"脊柱"干道一侧，一直延伸到室内的底层和顶层。为了与周围景观地形融为一体，该建筑"由内而发"，蜿蜒起伏，向外延伸。同时，建筑的内部开阔，人们很容易不期而遇；且学校活动不再被局限在有限的区域内，来鼓励合作、互动和社交生活。

munication technology and the significant mobility of students and teachers transcend the limitations of a situated institution. Students can have remote access to knowledge, complement or finish their studies in universities other than their own. Academics collaborate on an international basis and present their findings to the international audience through journals, conferences and teaching exchanges. Faculties collaborate on joint programs and federations. Institutions in developing countries or former colonies no longer send their students to the West, but recruit faculty and attract students from all over the world. Established institutions "franchise" and expand their presence outside their national borders. These are mere fractions of evidence to a global society – and perhaps even a global knowledge capital. Yet in spite of these developments we are still not far from the foundational model of a corporate activity, namely of "exchanging teachers, students and ideas through the lingua franca of Latin" for the universities in the Middle Ages, and the "cooperat[ion] in the recovery of ancient learning and the scientific discoveries of the Enlightenment" of the early modern ones[7].

What then of architecture? How does the production of building, extending to the idea of campus, serve this picture? For one thing we need to remember that architecture ultimately serves a purpose; yet, limiting itself to the mere facility of ergonomic requirements would unlikely serve the purpose well, as we have shown it to be far more than the mere function of teaching. Taking into consideration the framework we described, it becomes evident that the architecture bottom line is called upon to do what it does best: to establish place, to generate society, to excite or foster meaning. In order to turn space into a living situation, it pursues the abstract empowerment of a site rather than the formal manipulation of it. In this context the projects we examine negotiate and set sustainable terms with a local condition, be it the landscape, a relationship with the local community or a historically and semantically charged site. Furthermore, it is called upon to address a global audience, aiming to invite all that share the same aspirations into the universitas. In that context the projects we examine aim to manifest the identity and values of the institution, and to symbolize its stance with as clear an image as possible. Finally, in one of the most primal functions of architecture, it aims to shelter people and activity. The buildings we examine are designed to provide suitable spaces for classes, lectures, experiments, services, inasmuch as they are fostering community by introducing intermediate spaces, transcending rigid boundaries, and generating haphazard collisions.

由Rafael Hevia和Rodrigo Duque Motta设计的迭戈波塔利斯大学经济商贸学院看似是一个截然相反的案例。建筑群致力于表达强烈的身份特征，且由两个不同的塔楼组成，一个是正方体，一个是板状，位于下沉结构的两侧，而结构嵌入缓坡内。两座塔楼的设计语言互补，相互呼应，与四周沉闷的环境形成鲜明对比。不论是在楼上还是楼下，都能看到山坡和一系列的庭院。通过增建楼顶花园、几处露台和露天走廊，"日常生活与远处的地貌联系在一起"。

BUSarchitektur建筑事务所设计的奥地利维也纳金融大学（WU）校区的教学中心傲然挺立于众多明星楼宇之中，这个建筑群被戏称为"鸡尾酒会的嘉宾"[8]。该建筑群将奥地利维也纳金融大学分散的楼宇整合到一个校园，并且很好地融入城市中。这些建筑分布在一条长廊的两侧，既赏心悦目，又兼顾实用性。其中，教学中心独树一帜，外墙覆盖着暗橙色的考顿钢，使整座建筑像巨大的石柱，气势恢宏。这根"石柱"中间有裂缝，可以看到建筑内部开放的公共空间，并与外面的空地相连。教学中心向社区敞开大门，公众在全年的每一个时刻都可以参观外部和大楼一楼。

可以说，由Henning Larsen建筑师事务所设计的南丹麦大学科灵校区是各种张力的汇集点：首先校区是科灵的新地标；其次，校区后面是景色优美的河畔公园，前面是休闲广场，校区内景与它们产生对话，构成"内部相连的城市空间"；另外，校区还是低能耗设计，利用自然的不利因素，把它转化成可持续能源和环境控制。虽然目前还不是很明显，但形式上采用的三角形主题将屋顶、阳台、遮阳板、墙壁等各种因素转变成和谐统一的流线、环境和形象。相互作用的阴影、半通透性和色彩使独特的外观更加惹人喜爱，且具有明显的标志性，这个经典之作表达了创新精神和群体精神。

由Legorreta+Legorreta设计的卡塔尔乔治城涉外事务学院不仅有

The Abedian School of Architecture by CRAB Studio is located in Queensland, Australia. It is situated in the north-west quadrant of the campus of Bond University developing under a series of lily pad structures which shelter consecutive enclosures up to three levels high. The full volume of this space is run by a series of "scoops", namely concrete coves rhythmically attached to the side of the interior spine expanding all the way to the full height of the interior. The building develops "from within", sloping towards the outside almost in pursuit of dissolving into the landscape. At the same time, the interior of the building is open to encounters; it encourages collaboration, interaction and social life by refraining from limiting the school's activity to well defined boundaries.

The Economics and Business Faculty at Diego Portales University by Rafael Hevia and Rodrigo Duque Motta, seemingly, is a case of the exact opposite. Challenged with the task to articulate a strong identity, the complex consists of two distinctive towers, a cube and a slab, attached to the opposite sides of a sunken structure which is embedded in the sloping site. Sharing a complementary design language, the two expand as a dipole in dialogue with each other, yet with an unmistakable presence against an otherwise blunt surrounding. Nevertheless, the sloping landscape stays dominant on the eyes level with a series of courtyards, through and underneath the buildings. Along with the addition of a roof-garden, a number of terraces and its exposed corridors, "everyday life is connected to a distant geography".

The Teaching Center in WU Campus by BUSarchitektur stands proudly in a complex of star-studded buildings that has been likened to "guests at a cocktail party"[8]. The complex itself is the equivalent of Vienna University of Economics and Business(WU)'s consolidation of scattered buildings into a single campus, yet sustaining still a strong integration with the city. The buildings themselves act as veritable events to a promenade, featuring various formal responses to claims about their programmatic intentions. Among them, the Teaching Center owes its distinctive identity to the dark orange corten-steel cladding which gives it a monolithic appearance. This "monolith" is slit with ruptures which reveal open communal spaces in the interior of the building, as well as connect to the outside premises. In a generous offer to the community, the public is given access not only to the outside but also to the inside of the building on the ground level 24/7, 365 days a year.

University of Southern Denmark, Kolding Campus by Henning Larsen Architects could be argued to mark the spot for a fortunate convergence of various tensions; a landmark for Kolding, an "interconnected urban space" that puts its interior in dialogue with

222年的古老传统，还坚守学校"坚持国际参与"[9]的承诺，并且起到在两个相距11 000公里的校区之间协同的作用。学院的主要设计理念可能打破纪念碑式的建筑规模，强调开发小型部门，"更有家庭的感觉"。一系列的庭院和中庭体现同样的理念，提供更放松的氛围，更有亲切感，可以大大缓解在这所赫赫有名的高校学习的压力。最终形成一个综合体：即与当地环境融合的国际学校，以及与当地条件相一致的国际化建筑表达，非常符合学校"合二为一"的校训。

Stanton Williams建筑师事务所设计的伦敦艺术大学中央圣马丁艺术与设计学院的英王十字车站新校区的地理位置和维也纳金融大学新校区很相似。学校从原来位于伦敦中心的一座古建筑中搬出来，迁至英王十字车站区的一座改造的粮仓。"伦敦市中心的部分区域也注定要搬走。"Jane Rapley校长的讲话认为搬迁有好处。"我们处于交通枢纽……不仅是进来的问题……还关系到走出去，关系到联系世界。"[10]校区的各个场所，穿过中心流线轴的悬浮的小径，教室、院系、建筑之间灵活的关系，现存和新材料之间的对话，都体现了转型的理念。这所建筑的设计为室内的人留出空间，以符合建造目标。

由HASSELL与Richard Kirk建筑师事务所联合设计的昆士兰大学现代工程学院大楼，位于布里斯班，该建筑确确实实地体现了使学习便利化的目标。该建筑不仅特征很有吸引力，而且还依据创新的可持续发展策略，充分利用布里斯班温和的亚热带气候。自然通风符合就地取材和处理的可再生资源的需求，这些资源包括用于结构和立面的胶合木材，而南面独特的赤褐色遮阳板不仅能遮挡阳光，还能随着光线的移动展现有趣的动画效果，成为交界面。整个空间被巧妙地分成不同的形状和高度，使整座建筑物好像从周围景观演变而来。同时，与建筑物等高的中央脊柱似乎是自然的一部分，源自大楼周围的自然环境，又与之融为一体。

both a scenic park on the adjacent river at the back and a recreational town plaza at the front, a low-energy design that harvests what ought to be natural impediments and turns it in sustainable energy and climate control. Although not apparent right away, its formal manipulations of a triangular theme turn its various elements – roofs, balconies, shades, walls – into a sweeping fugue of circulation, atmosphere and imagery. The soft interplay of shade, semi-transparency and color, energizes its distinctive facade in a welcoming, yet unmistakably iconic, expression of creativeness, and community.

Georgetown School of Foreign Service in Qatar by Legorreta + Legorreta carries not only a 222 years old tradition but also the university's commitment to its "tradition of international engagement"[9], acting as the vehicle of synergies between two campuses that stand "eleven thousand kilometers away". Not unlikely then, the principal design principle was to eschew a monumental scale and focus instead on the development of smaller departments to "give the idea of being at home". The same principle is manifested with a series of courtyards and atriums which provide for a more relaxing atmosphere and a sense of intimacy invaluable for decomposing the pressure of studying at a highly prestigious institution. The end result is a form of synthesis; an international school in a local context and an international expression of architecture in according with a local condition, quite consistent to the university's motto "utraque unum" (both into one).

The New University of the Arts London Campus for Central Saint Martins at King's Cross by Stanton Williams Architects enjoys a locality not unlike to the new WU campus in Vienna. The School moved from its former premises on a heritage building at Central London to a converted granary at King's Cross, "where part of the shift of the center of London was bound to move". The words of Head of College Jane Rapley deem this move beneficial: "we're in this transport hub (…) it's not about just coming in, but coming out (…) it's about connecting with the world"[10]. The notion of transitions is evident within the premises of the complex, with suspended walkways crossing the central circulation spine, as well as with the flexible relationships that are developed between classrooms, different departments or buildings, or the dialogue between existing and new materials. This building is designed to leave room for the people inside to match its ambition.

The University of Queensland Advanced Engineering Building in Brisbane by Richard Kirk Architect and HASSELL is a building that literally embodies the very subject of study it aims to facilitate. Apart from its appealing character, the building is organized

最后，由Heatherwick工作室设计的新加坡南洋理工大学学习中心重新定义了现有环境。它的建造灵感是把许多独有的空间（学习空间或者简单地说教室）集中起来，使它们聚集、叠加，变成排列在开放且通透的中庭周围的凹形空间，而中庭的设计符合当地气候条件和文化，可以让光线射进来，维持空气循环。Heatherwick工作室突破常规的大学建筑形式，摒弃了笔直的大型走廊和整齐排列的无名教室。这种设计有一种无形的优点，可以“促进聚会和社交”[11]。人们根据情况参与活动，结成小组，在中庭和阳台的开放空间会面，在数字空间获取教学和研究资料。从外观看，整体自然地演变成一个体量，外观与南洋理工大学其他“常规”建筑形成鲜明对比。而最与众不同的措施是通向一层的通道不必再分等级，整栋大楼360度全开放，学生可以自由进出。

around innovative sustainability strategies taking advantage of the benign sub-tropical climate of Brisbane. Natural ventilation meets with locally sourced and treated renewable resources such as gulam timber for the structure and facade, while the distinctive terracotta shade on the south not only acts its purpose but serves as an interface for its playful animation with the passing of the light. With the clever break of its volume into parts of different shapes and heights, the building seems to evolve from the landscape; at the same time a full height central spine seems to be deriving as a natural sequence from – and towards – the natural surroundings of the building.

Last, Heatherwick Studio's Learning Hub at the Nanyang Technological University in Singapore, stands mostly as a redefinition of an existing context. For one thing its very form is built up from the idea of collecting a number of singularities – learning spaces, or, in plain talk, classrooms – into a stacked accumulation of coves organized around an open-air central atrium, an arrangement consistent with local climate and culture which lets light and natural air in. Heatherwick Studio was set to do away with large linear corridors and the nameless, orderly arranged classrooms that are the norm in university buildings. This arrangement was substituted with an intangible quality which aims to "foster togetherness and sociability"[11]: people organize in groups formed by occasion, they meet in the abstract openness of the atrium standing in balconies, and they access teaching and research material in digital space. On the outside the lot evolves organically into a volume whose appearance is in distinct contrast to its backdrop, formed by the other "regular" buildings of the NTU. In an ultimate act of redefining the ordinary, the building eschews the hierarchy of organized access to its ground level by opening up 360° around its large central core. Angelos Psilopoulos

1. Jacques Verger, "Patterns", A History of the University in Europe: Volume 1. Universities in the Middle Ages, ed. Hilde de Ridder-Symoens, Vol. 1 (Cambridge University Press, 2003), p.37~38.
2. "Academy", *Oxford Dictionaries* (Oxford University Press, 2015), <www.oxforddictionaries.com>
3. Harold Perkin, "History of Universities", *International Handbook of Higher Education*, ed. James J. F. Forest and Director Philip G. Altbach, Springer International Handbooks of Education 18 (Springer Netherlands, 2007), p.159, <http://link.springer.com/chapter/10.1007/978-1-4020-4012-2_10>
4. in the context of social, cultural, historical, political, geographical (...) etc. relations, see Daniel Chandler and Rod Munday, eds., "Situatedness", *A Dictionary of Media and Communication* (Oxford University Press, 2011), <www.oxfordreference.com>; Also, Phillip Vannini, "Situatedness", *The SAGE Encyclopedia of Qualitative Research Methods* (CA: SAGE Publications, 2008), <http://knowledge.sagepub.com>
5. Frederic Duncalf and August C. Krey, *Parallel Source Problems in Medieval History* (New York, London: Harper & Brothers, 1912), p.137~141, <https://openlibrary.org>
6. Harold Perkin, "History of Universities," p.192.
7. Ibid., p.196.
8. Charlotte Skene Catling, "Sextet in the City: Vienna University's Starchitect Campus", The Architectural Review, May 27, 2014, <www.architectural-review.com>
9. "Remarks by President John J. DeGioia - Building Inauguration at the School of Foreign Service in Qatar", Georgetown University, February 13, 2011, <www.georgetown.edu>
10. *Central Saint Martins, New King's Cross Campus*, 2011, <http://youtu.be/ZXNrzzi_S2g>
11. "Learning Hub|Heatherwick Studio", accessed March 13, 2015, <www.heatherwick.com>

新加坡南洋理工大学学习中心

Heatherwick Studio

新加坡南洋理工大学学习中心是新加坡全新的教育建筑地标。作为南洋理工大学校园重建规划的一部分，这座学习中心被设计成一栋全新的多功能建筑，服务于33 000位学生。该建筑突破了传统教育建筑以长廊连接盒状教室的设计格局，追求更为优越、更适于现代学习模式的独特设计。数字革命使教学活动可以在任何地点展开，该建筑最重要的功能就是为各个专业的学生和教师提供聚集和相互交流的场所。期望在这里学生们能够遇到他们未来的生意合作伙伴，或者志同道合的朋友。

该建筑融社交与学习空间为一体，打造更加适合学生与教师之间的非正式或即时交流互动的动态环境。建筑包括十二座塔楼，每座塔楼由圆形辅导室叠加组成，环绕中部的宽敞中庭向底部逐渐收拢，形成五十六间没有拐角、没有明显前后之分的教室。

这种新一代的智能教室设计，能够提升分组教学过程中的交互性，有利于学生的自主学习，因此为南洋理工大学新型教学法的实施提供了支撑。灵活的教室布局既方便教师们按照教学需要组织安排学生，也为学生之间更好的相互合作提供便利。

围绕中庭的空间设置成公共流线，教室入口面向流线次第敞开。公共空间和不规则布局的花园露台在整栋建筑中随处可见，为学生创造视觉交互的机会，提供驻足、聚集、小憩的空间。

无论是新加坡的地方建筑法规，还是建筑本身所承载的高环保期望值，都决定了本案需要采用混凝土结构。在建筑设计过程中，遇到的首要难题就是如何运用这种质朴的材料呈现出美丽的建筑图景。

混凝土楼梯与电梯外墙立体浇铸了700幅特别设计的图画，内容涉及科学、艺术、文学等各种主题。这些重叠交错的图案由特别委托的插图画家Sara Fanelli设计完成，意在激发人们的模糊思维，留下想象力发挥的空间。61根倾斜角度各异的混凝土立柱为本案勾勒出独特的波浪结构。曲形的立面板材上铸有特别的水平纹理，这些纹理由10个可调节的硅树脂磨具制作而成，以经济实惠的方式打造出复杂的三维立体结构。项目采用各式各样的原始手法对混凝土材料进行处理，使整栋建筑呈现出粘土手工制作的效果。

由于新加坡的气温常年保持在25到31摄氏度之间，因此在确保使用可持续性能源的同时满足学生对室温舒适感的需求就显得尤为重要。

建筑开放、通透的中庭实现了自然通风效果，保证环绕塔楼教室的空气得到最大程度的循环流通，尽可能使学生感觉凉爽舒适。每间教室都设有静音制冷装置，避免人们对高能耗空调扇的依赖。学习中心大楼获得新加坡建设局颁发的绿色建筑标志白金奖，该奖项也有此类建筑所能达到的最高级别的环境标准。

AHEAD

Learning Hub in NTU

Nanyang Technological University(NTU Singapore) Learning Hub is a new educational landmark for Singapore. As part of NTU's redevelopment plan for the campus, the Learning Hub is designed to be a new multi-use building for its 33,000 students. Instead of the traditional format of an educational building with miles of corridors linking box-like lecture rooms, the university asked for a unique design better suited to contemporary ways of learning. With the digital revolution allowing learning to take place almost anywhere, the most important function of this new university building was to be a place where students and professors from various disciplines could meet and interact with one other. The Learning Hub is envisioned to be a place where students might meet their future business partner or someone they would have an amazing idea with.

The outcome is a structure that interweaves both social and learning spaces to create a dynamic environment more conducive to casual and incidental interaction between students and professors. Twelve towers, each a stack of rounded tutorial rooms, taper inwards at their base around a generous public central atrium to provide fifty-six tutorial rooms without corners or obvious fronts or backs.

The new-generation smart classrooms were conceived by NTU to support its new learning pedagogies that promote more interactive small group teaching and active learning. The flexible format of the rooms allows professors to configure them to better engage their students, and for students to more easily collaborate with each other.

The rooms in turn open onto the shared circulation space around the atrium, interspersed with open spaces and informal garden terraces, allowing students to be visually connected while also leaving space to linger, gather and pause.

The combination of local building codes and high environmental aspirations meant that a concrete construction was necessary. The primary design challenge was how to make this humble material feel beautiful.

The concrete stair and elevator cores have been embedded with 700 specially commissioned drawings, three-dimensionally cast into the concrete, referencing everything from science to art and literature. Overlapping images, specially commissioned from illustrator Sara Fanelli, are deliberately ambiguous thought triggers, designed to leave space for the imagination. The sixty one angled concrete columns have a distinctive undulating texture developed specially for the project. The curved facade panels are cast with a unique horizontal pattern, made with ten cost-efficient adjustable silicone molds, to create a complex three-dimensional texture. The result of the building's various raw treatments of concrete is that the whole project appears to have been handmade from wet clay.

With year-round temperatures in Singapore between 25°C and 31°C it was important to maintain the students' comfort whilst achieving a sustainable energy usage.

The building's open and permeable atrium is naturally ventilated, maximizing air circulation around the towers of tutorial rooms and allowing students to feel as cool as possible. Each room is cooled using silent convection, which does away with the need for energy-heavy air conditioning fans. The Learning Hub building was awarded Green Mark Platinum status by the Building and Construction Authority (BCA), Singapore, the highest possible environmental standard for a building of this type.

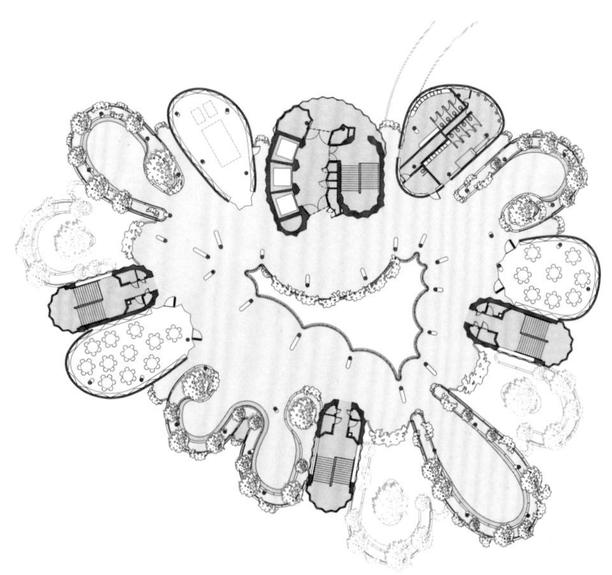

八层 eighth floor

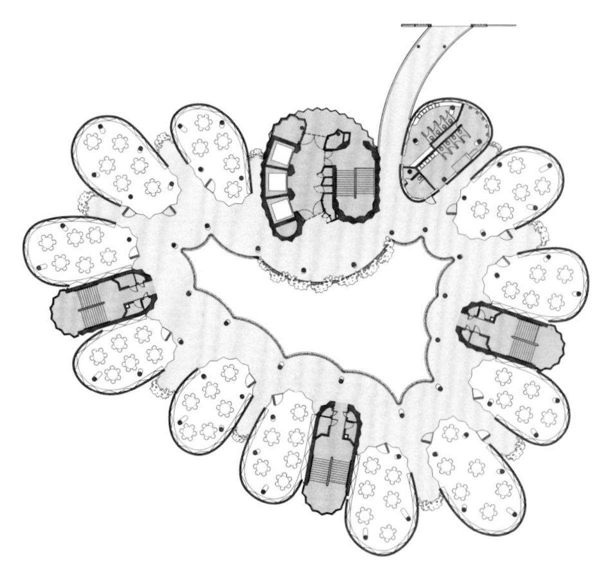

四层 fourth floor

项目名称：Learning Hub
地点：Nanyang Technological University, Singapore
建筑师：Heatherwick Studio
可持续性顾问：CPG Consultants
M&E工程师：Bescon Consulting Engineers
土木&结构工程师：TYLin International
环境设计：Green Mark Platinum
用地面积：2,000m² / 有效楼层面积：14,000m²
建筑高度：38.3m / 建筑规模：8 storeys
施工时间：2012.8 / 竣工时间：2015.3
摄影师：©Hufton + Crow (courtesy of the architect)

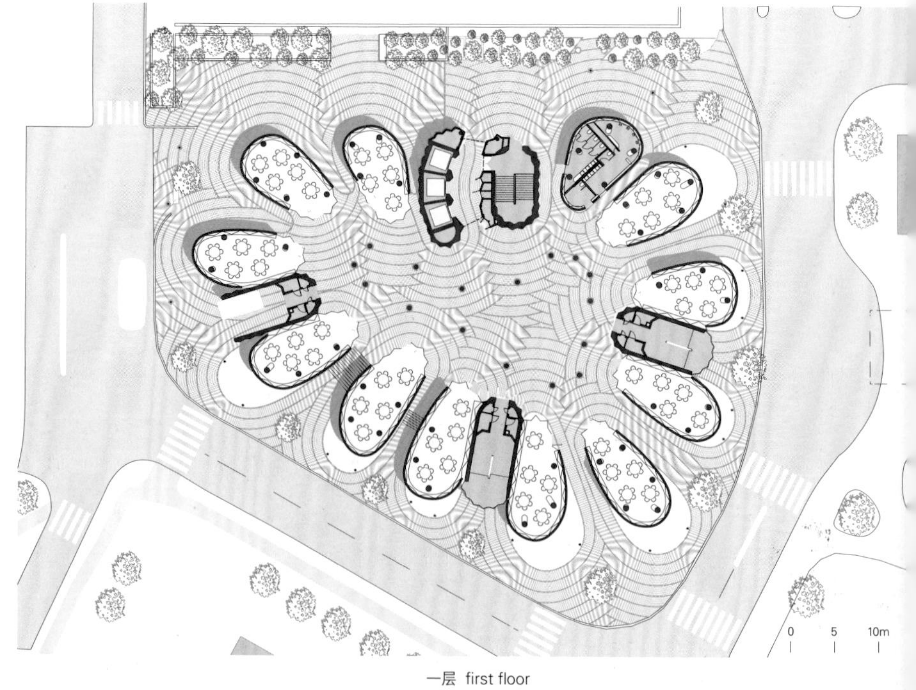

一层 first floor

EXIT

邦德大学Abedian建筑学院

CRAB Studio

亲临邦德大学新建的Abedian建筑学院，犹如体验一次情节多样的旅程。建筑的北立面进行遮阳处理，南向风格则轻快、开放，且通风良好。建筑内部弯曲的主干道成为校园西北区域的软核心，这里充满了学生实验、社会活动、学术讲座、观点评论以及奇闻趣事等学院生活的气氛。

在这条主干道之外，一条宽阔的室内步道从安静的学习区（"鼻状"凸起部分的下方）穿过，串联起若干个"勺形"空间。通过内部蜿蜒的"山丘"，建筑结构形成上升与下落的态势，与周边的地势相互呼应。沿着这条"街道"，教师工作室和大型集会空间在阶梯式平台上分布开来，平台自身也融入一座重新种植的山坡花园之中。这些"建筑板块"与中央"干道"、安静的学习区、引人注目的"勺形"空间，共同定义了该建筑的硬核心。与之形成良好对照的是，大部分的滤光"表皮"和"嵌入结构"具有更为柔软、更加透明的特点。

本案营造出一种轻松的氛围，每个人都能在此找到非常适合其活动的场地。尽管工作室区域、勺形区域、平台和角落的设计都基于一个鲜明的层级和体系，但是这些空间的方位、规模都有非常大的变化。建筑师希望这些设计元素能够以一种微妙且令人愉悦的方式独树一帜。

建筑师还期望巧妙地处理建筑的外观，使其遮蔽北向的直射阳光，并将南向的滤光引入建筑内部，既防止眩光和室内过热，也避免了室内空间体验的同质性，因此建筑师设计了一系列各具特色的"明亮空间"和相对较暗的、较基础的凹型空间。设计方案利用建筑外围护保温结构，将多余的太阳能获取量降至最少。

屋顶和立面系统的设计预先考虑到昆士兰州夏季阳光可能达到的强度和照射的方位。建筑各个洞口方向的设计、遮阳屋盖以及立面的立

柱系统成功地降低了过度日照可能造成的影响。

建筑的朝向设计减少了北向、东向以及西向墙体和窗户可能的暴露面。在这些方向的立面上，每处洞口都避开了强烈的日光直射。所有北向窗户都另外安装了夏季遮阳装置。在面积最大的西立面，外部的垂直立柱和悬挑屋顶为工作区域遮挡了夏季阳光强烈的间接照射。

在建筑设计方案的开发阶段，该地区的风力风向情况也被纳入考虑范畴。建筑环抱山顶，以其坚实的"背部"抵挡寒冷的南风。北向立面设置规格各异的窗户，利用一系列金属"眉状"结构来遮挡阳光的侵袭，这些眉状结构在形成一件艺术作品的同时，也形成一个有效的遮蔽系统。

建筑内部沿中央"街道"形成主要的空气流通区域，该区域能够促进建筑内部在长度方向的自然通风，因此具有热缓冲器的作用。

Abedian School of Architecture at Bond University

Bond University's new Abedian School of Architecture might be experienced as varied and episodic journey. Sheltered and determined to the north, the building is airy, effortless and free to the south. The curvature of its spinal interior route establishes a new soft core for the Northwest Quadrant of the Campus – a core populated by the life of the school, by student experimentation, social gatherings, lectures, critics and weird happenings.

Leaving the existing spine pathway, the broad internal path dives underneath the nose of the quiet-study strip and proceeds past a series of "scoops". Via its meandering internal "hill", the rise and fall

北立面 north elevation

of the building gently echo the topography of its surroundings. From this "street" the faculty's studios and large gathering spaces spread out onto a terraced deck – which itself melts into a re-vegetated hillside garden. These "tectonic rafts" together with the central "spine", quieter study areas and dramatic "scoops" define the building's rocky core. In welcomed contrast a majority of the filtering "wraps" and "insertions" are of a softer, more translucent character.

The architects created a very ambient building, where the individual can really identify with the nature of his or her activity – thus the studio pads, scoops, decks and corners – though based on a clear hierarchy and system – have significant shifts of direction or variations of size. These elements particularize – the architects hope, in a subtle and enjoyable way.

The architects wished to manipulate the surface of the building – sheltering it from direct northern light and filtering the southern

照片提供：©Andy Wingate (courtesy of the architect)

南立面 south elevation

东立面 east elevation

light into the interior, avoiding glare and overheating – without homogenizing one's experience the interior. The architects created instead, an idiosyncratic series of "lit places" and darker, more elemental pockets.

This proposal is designed from the ground up to minimize undesirable solar gain within the building's thermal envelope.

Both the roof and facade systems anticipate the potential strength and direction of the Queensland summer sun. The orientation of the building's openings, together with the sunhoods and column system of the facades succeeds in mitigating a majority of the sun's potentially excessive effects.

The building's orientation reduces the potential exposure of the north, east and west facing walls and windows. On these facades, each opening is sheltered from the strongest direct sunlight. Additional summer solar protection has been applied to all north facing windows. On the largest, west elevation external vertical columns and overhanging roof protect the studio areas from the harshest indirect summer sun.

The architects have also taken account of the prevailing winds during the development stage of the scheme. The building encircles the crown of the hill, turning its harder "back" to the colder south winds, whereas the north facade shelters a variety of window configurations that are screened from the sun by a series of metallic "eyebrows" that effectively create an art piece as well as a shielding system.

Internally, the main circulation areas along the internal "street" act as a thermal buffer and encourage the natural movement of air along the length of the building. CRAB Studio

Abedian
School of Architecture

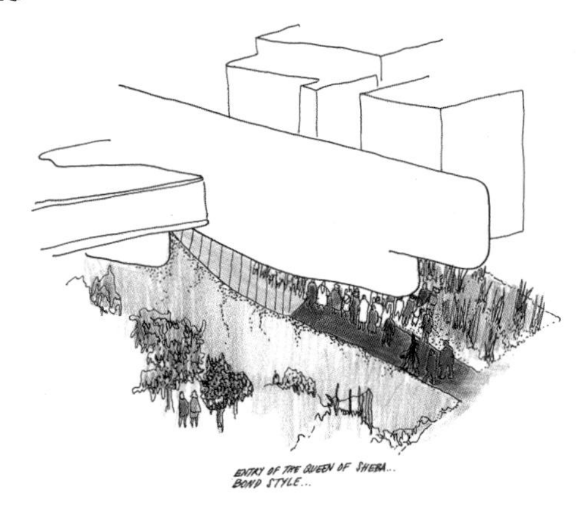
ENTRY OF THE QUEEN OF SHEBA...
BOND STYLE...

THINGS WEREN'T GOING
SO WELL
THAT NIGHT....

SHIT! I HOPE THE CRAYFISH
IS STILL 'ON'.....

MY GOODNESS, KATE
IT'S ALL HAPPENING!

THESE GUYS INSIST THAT
ARCHITECTURE IS AN ART FORM

LOOK GUYS......
YOU HAVE TO ACCEPT IT
AS A POINT OF DEPARTURE

项目名称：Abedian School of Architecture / 地点：Queensland, Australia
建筑师：Cook Robotham Architectural Bureau Ltd.
项目团队：Peter Cook, Gavin Robotham, Mark Bagguley, Jenna Al-Ali, Ting-Na Chen, Lorene Faure, Yang Yu, Tim Culverhouse
结构/机械和电气/音效工程师：Arup Pty Ltd.
总承包商 / 项目管理：ADCO Constructions Pty Ltd.
甲方：Bond University
用地面积：9,000m² / 总建筑面积：2,500m²
造价：AUD 16.2m
竞赛时间：2010.11 / 施工时间：2012.1 / 竣工时间：2013
摄影师：
©Peter Bennetts(courtesy of the architect)-p.18~19, p.22~23, p.25, p.27, p.30
©Rix Ryan Photography(courtesy of the architect)-p.20~21, p.26~27, p.28, p.29, p.32top, p.33, p.34~35

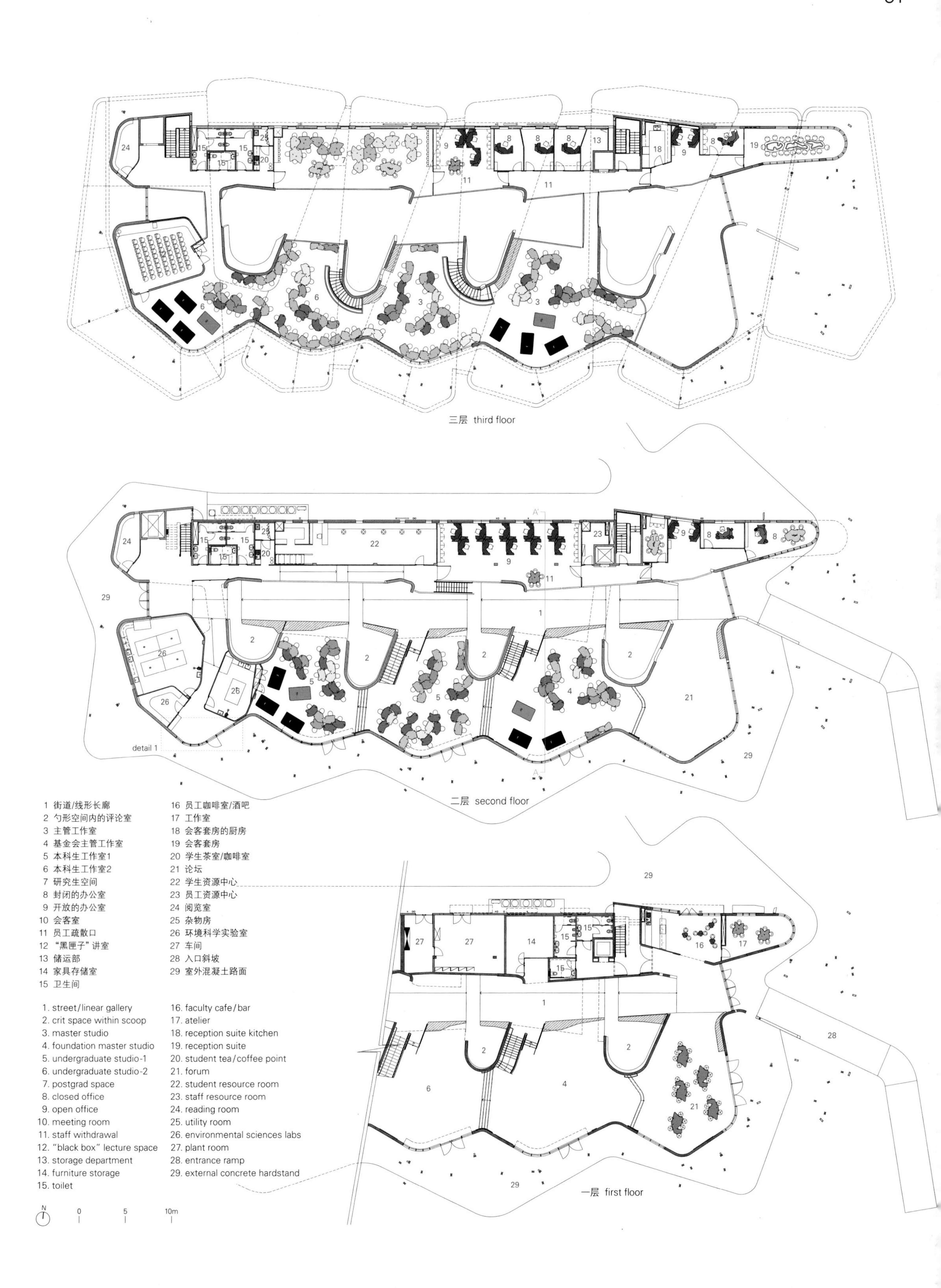
三层 third floor
二层 second floor
一层 first floor
detail 1
1 街道/线形长廊
2 勺形空间内的评论室
3 主管工作室
4 基金会主管工作室
5 本科生工作室1
6 本科生工作室2
7 研究生空间
8 封闭的办公室
9 开放的办公室
10 会客室
11 员工疏散口
12 “黑匣子”讲室
13 储运部
14 家具存储室
15 卫生间
16 员工咖啡室/酒吧
17 工作室
18 会客套房的厨房
19 会客套房
20 学生茶室/咖啡室
21 论坛
22 学生资源中心
23 员工资源中心
24 阅览室
25 杂物房
26 环境科学实验室
27 车间
28 入口斜坡
29 室外混凝土路面
1. street/linear gallery
2. crit space within scoop
3. master studio
4. foundation master studio
5. undergraduate studio-1
6. undergraduate studio-2
7. postgrad space
8. closed office
9. open office
10. meeting room
11. staff withdrawal
12. “black box” lecture space
13. storage department
14. furniture storage
15. toilet
16. faculty cafe/bar
17. atelier
18. reception suite kitchen
19. reception suite
20. student tea/coffee point
21. forum
22. student resource room
23. staff resource room
24. reading room
25. utility room
26. environmental sciences labs
27. plant room
28. entrance ramp
29. external concrete hardstand
N
0
5
10m

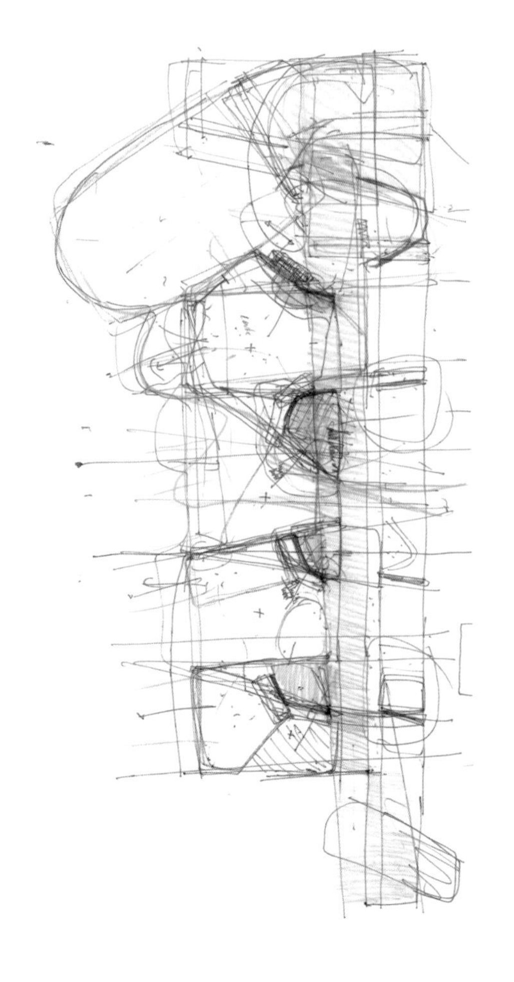

照片提供：courtesy of the architect

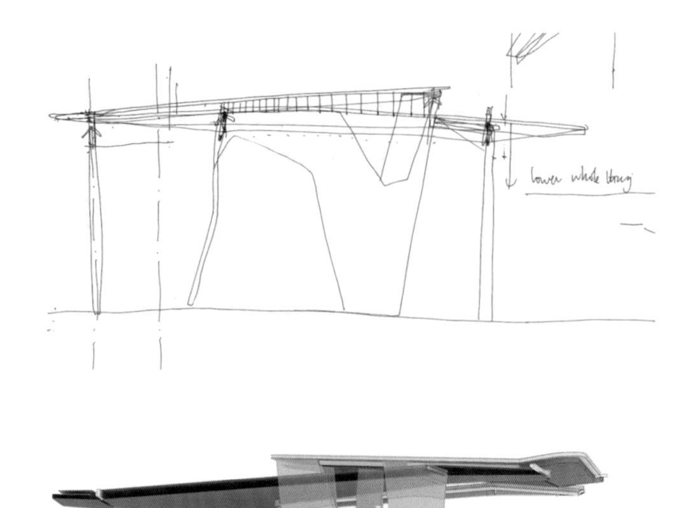

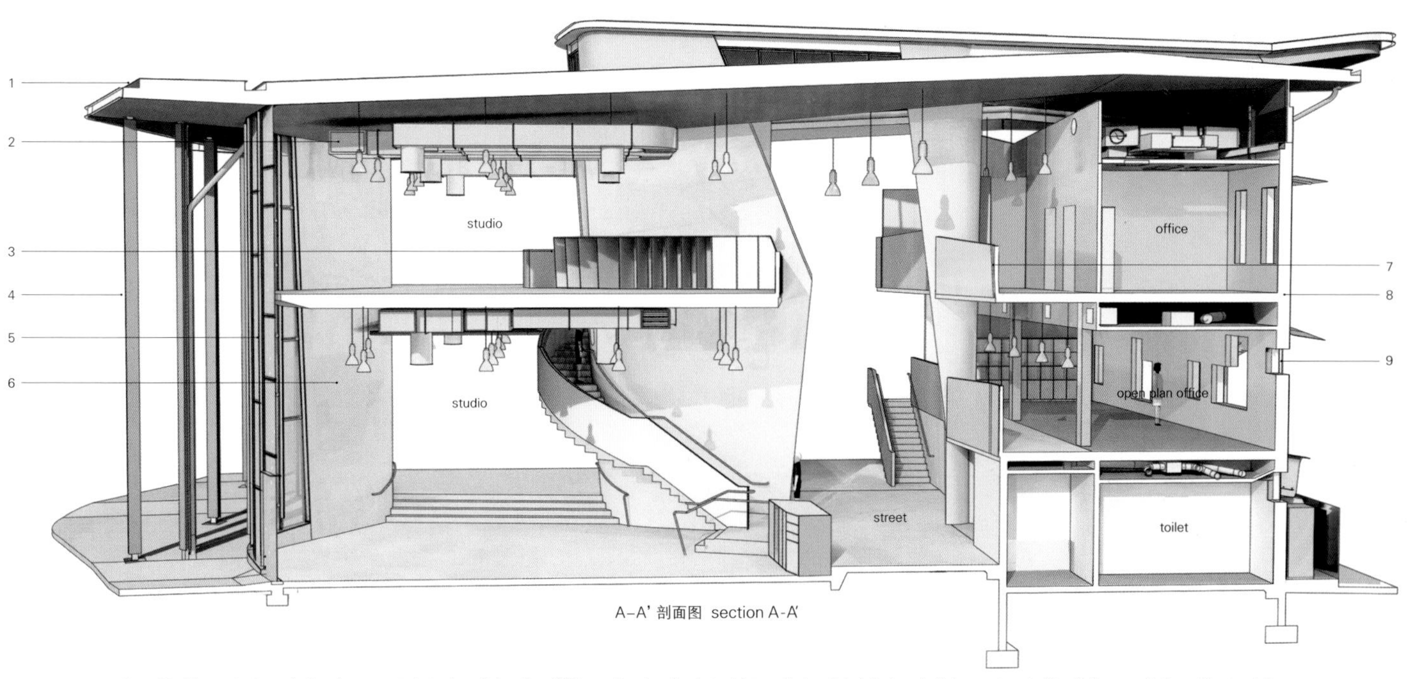

A-A' 剖面图 section A-A'

1. roof build up: sloping colorbond corrugated steel roof sheeting/200mm 2. galvanised steel internally insulated ductwork 3. hoop pine ply fitted joinery unit 4. architectural fin: 150x100mm rectangular hollow section with 300x35 mm timber cladding 5. aluminium composite window: 230mm aluminium plate mullion and sill/160mm transoms/structural silicone glazing with dow corning 795 black silicone/low emissivity heat strengthened laminated glass 6. south facade build up: 250mm reinforced concrete wall 7. ply balustrade build up: 50x65mm SHS posts with 18mm hoop pine ply cladding 8. north facade build up: 6 mm render and paint/190mm blockwork/25mm cavity/100mm insulation/18mm hoop pine ply 9. aluminium window frame: 100x60mm aluminium framing system/low emissivity heat strengthened laminated glass

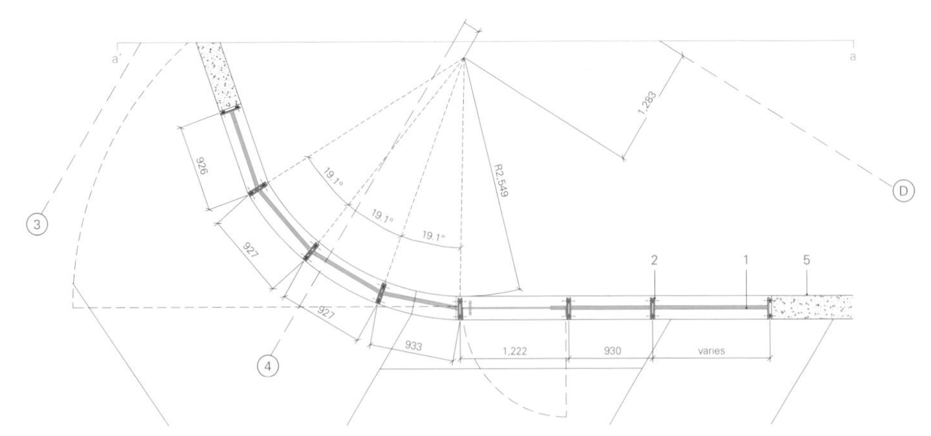

详图1 detail 1

1. structural silicone glazing with dow corning 795 black silicone/low emissivity heat strengthened laminated glass
2. mullion assembly: 16mm x 30mm (X2) aluminium plate
3. frameless glazed door
4. 150mm concrete slab
5. 250mm concrete wall
6. transom 16mm x 160mm aluminium plate
7. 325mm concrete slab
8. 20mm step down at door thresholds 30mm fall to threshold ramp in accordance

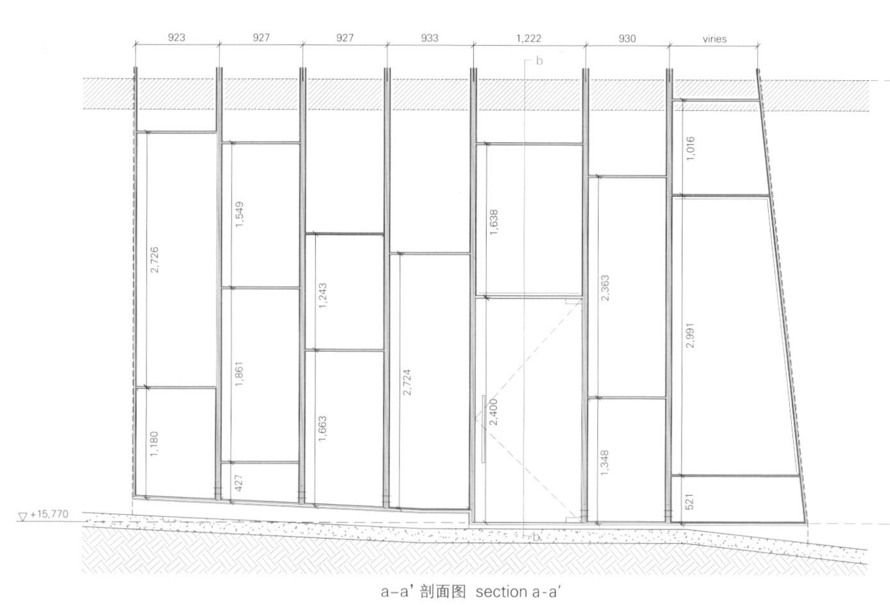

a-a' 剖面图 section a-a'

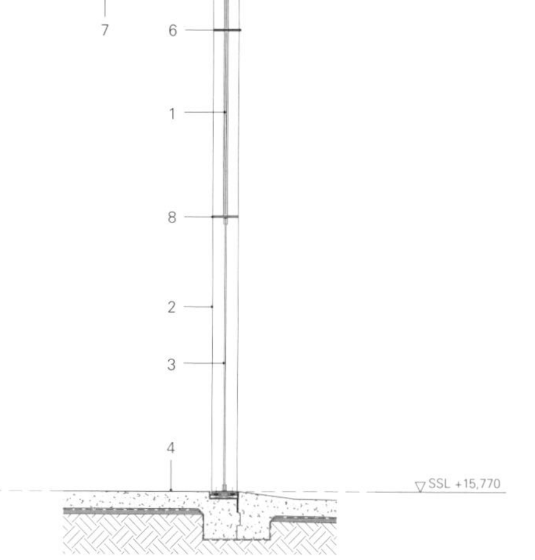

b-b' 剖面图 section b-b'

迭戈波塔利斯大学经济商贸学院

Rafael Hevia + Rodrigo Duque Motta

该项目是迭戈波塔利斯大学经济商贸学院的Huechuraba新校区翻新规划设计的获胜作品，同时也是该项目的一期项目。

建筑场地位于圣克里斯托瓦尔山脚下，Huechuraba山谷里倾斜且略微上升的地势之上。该地区是城市中最重要的商业中心之一，而新校区选址表示学院意在寻求将学术发展与专业实践紧密结合。

从建筑学的角度来看，这代表了以一种学术性的身份进驻到一处被完全不同的规则所主导的环境之中所要面临的挑战。周围的建筑多为租赁办公室，均为清一色的"玻璃盒子"，缺乏明显特征与设计逻辑。该项目意图打造一座与众不同的建筑，一个富有重量感的结构，持久而稳定，与大学所承载的义务和展现的卓越性一脉相承。

建筑体量的密度，经久耐用的混凝土，覆满葡萄藤、反映季节变化的墙壁，常年开放的公园和石造庭院，共同营造出一处历久弥新的空间。此外，该项目寻求利用地理环境来与校外的邻居建立联系。

庭院利用斜坡地势的优势，占据了整片区域，此外这里还建有高低不同的露台和一座屋顶花园，将建筑内进行的日常生活与远处圣克里斯托瓦尔山和Huechuraba山谷联系。这是将休闲时间与周围景观紧密结合的场所。

该项目是大学重新建造可持续发展环境设施的强烈愿望之一。项目设计从建筑、景观和技术设施的角度，以节能节水的形式，将对环境的影响降到最低，确保为使用者提供最舒适的环境。建筑寻求降低太阳能对建筑外壳的影响：外墙布满高度可控的洞口，以提供足够良好的采光和视野，中空空间采用间接照明，流线围绕其进行布局。此外，设计方案提议在北侧和西侧的外立面安装植被系统，落叶藤蔓能够在气候炎热的月份为人们提供荫蔽和湿气，而一个屋顶花园则充当了抵挡热气流的缓冲区。

在大学生教学楼，东侧外立面由一面水泥墙改建而成，外层覆以装饰玻璃，倒映着山影，并通过可控的洞口过滤射入教室的光线。西侧，一个巨大的垂直混凝土格架设有种植了多年生植物的花槽，为走廊和教室遮阴。最后，建筑寻求在气候炎热的月份开发通风系统，在教学建筑中实现对流通风，并在研究生教学楼的中空区域实现气流上升运动。

Economics and Business Faculty at Diego Portales University

The project is the result of the competition for the revision of the new Huechuraba Campus' master plan, and the definition of the project for its first phase, the Economics and Business Faculty of the Diego Portales University.

The site is located at the foot of San Cristobal Hill, sloping and slightly raised above the Huechuraba Valley. With the location of the new campus, the Faculty seeks to build a strong link between its academic development and the professional reality, as it is at one of the most important business centers in the city.

From the architectural point of view, this presented the challenge of asserting an academic identity in an environment governed by very different rules. Amid buildings that are most for office rental, and are glass boxes lacking a clear identity and designed with a short term logic, the project wanted to build a contrast, a structure with weight that speaks of permanence and stability,

udp
797

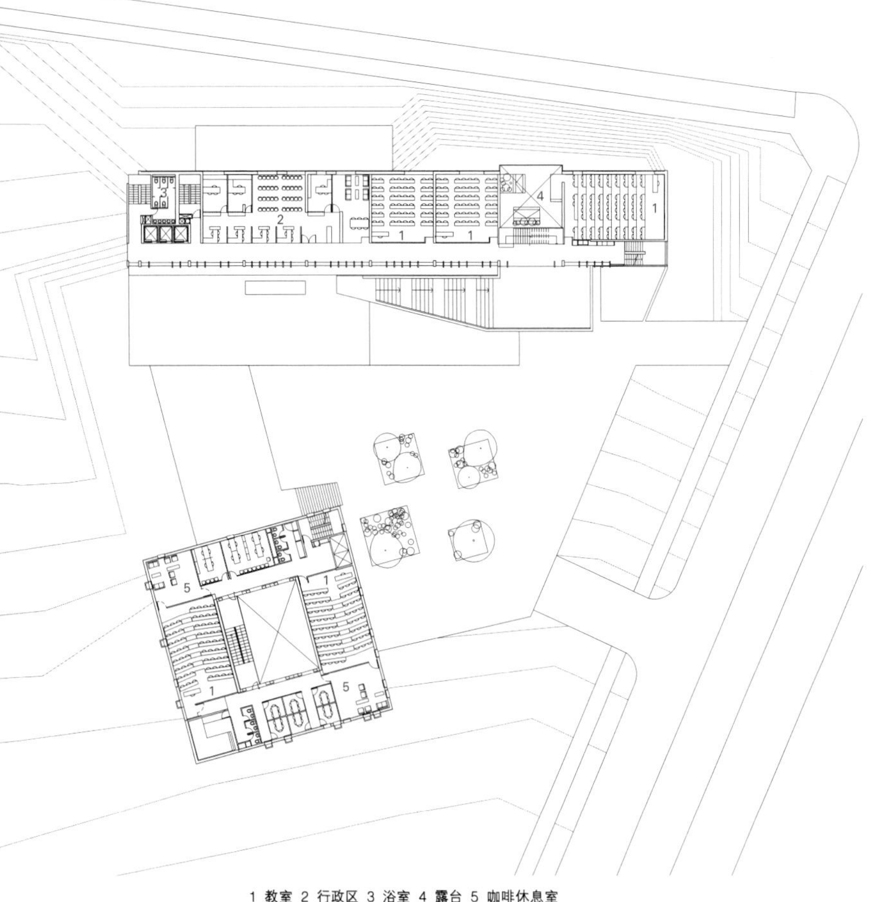

1 教室 2 行政区 3 浴室 4 露台 5 咖啡休息室
1. classroom 2. administration 3. bathroom 4. terrace 5. coffee break
二层 second floor

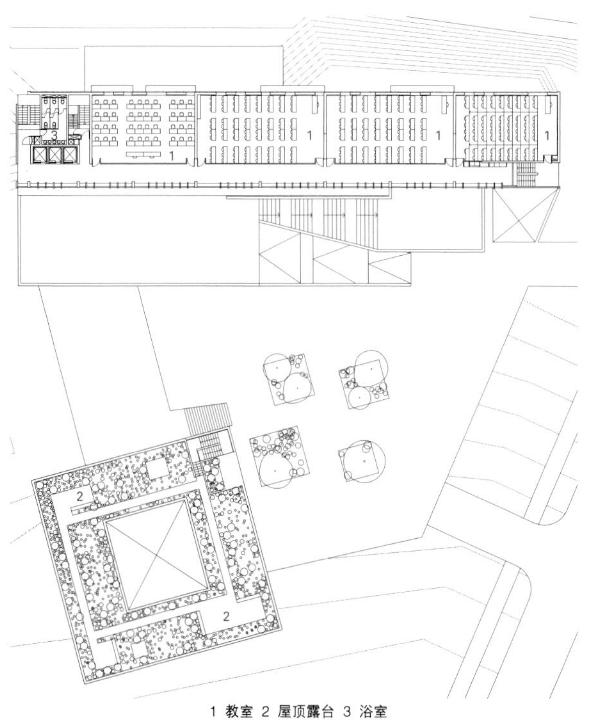

1 教室 2 屋顶露台 3 浴室
1. classroom 2. roof top terrace 3. bathroom
八层 eighth floor

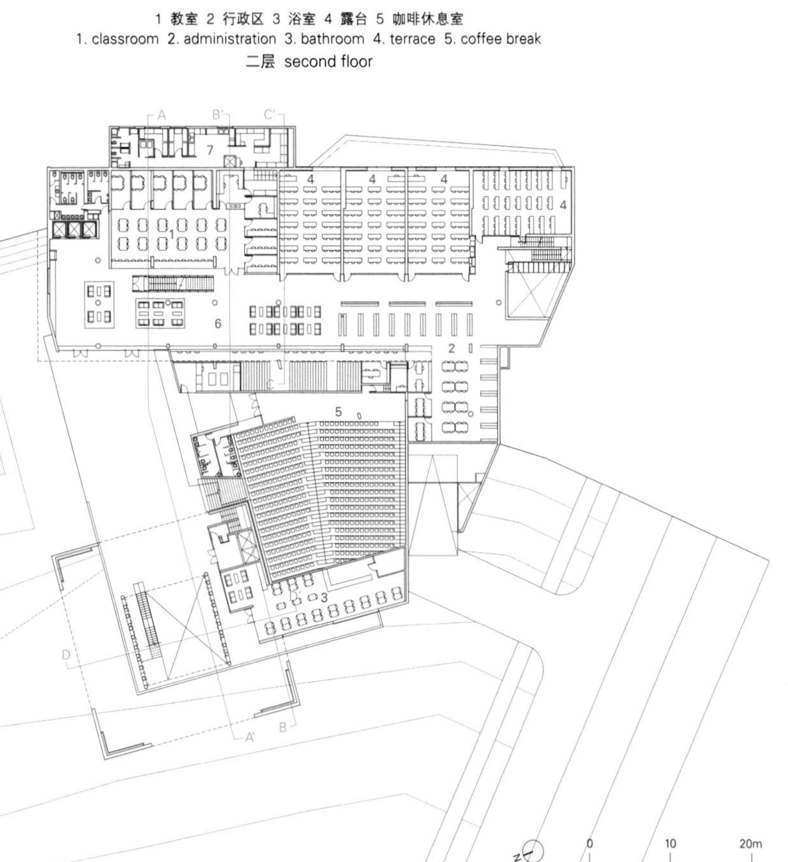

1 自习室 2 图书馆 3 自助餐厅 4 计算机实验室 5 礼堂 6 阅览区 7 厨房
1. study room 2. library 3. cafeteria 4. computer lab 5. auditorium 6. reading area 7. kitchen
地下一层 first floor below ground

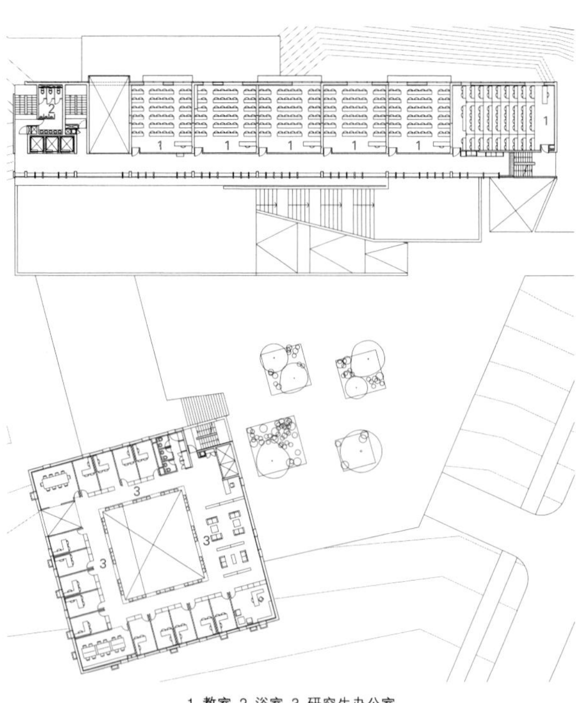

1 教室 2 浴室 3 研究生办公室
1. classroom 2. bathroom 3. graduate office
五层 fifth floor

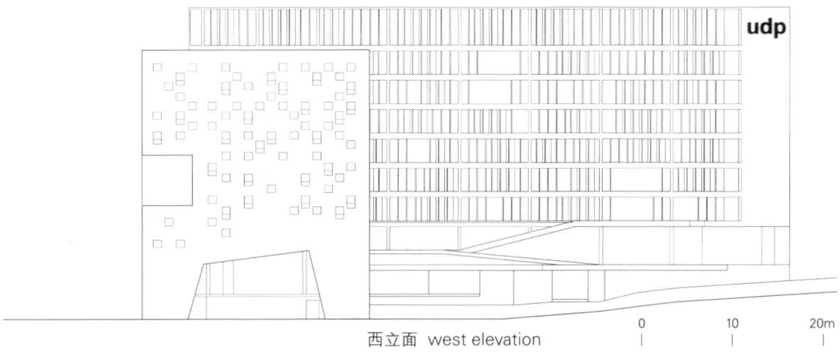

西立面 west elevation

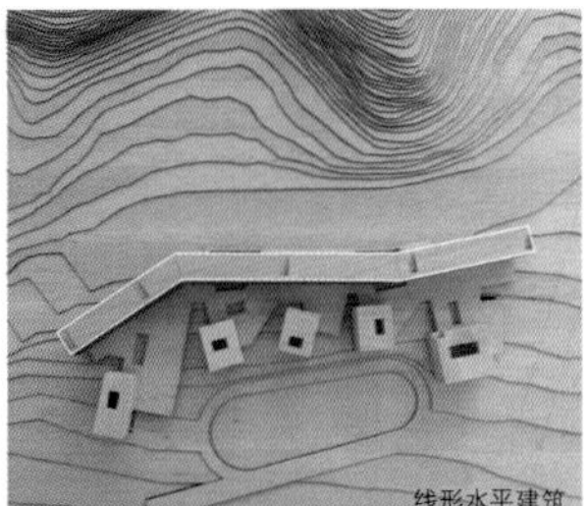

北立面 north elevation

线形水平建筑
long horizontal building

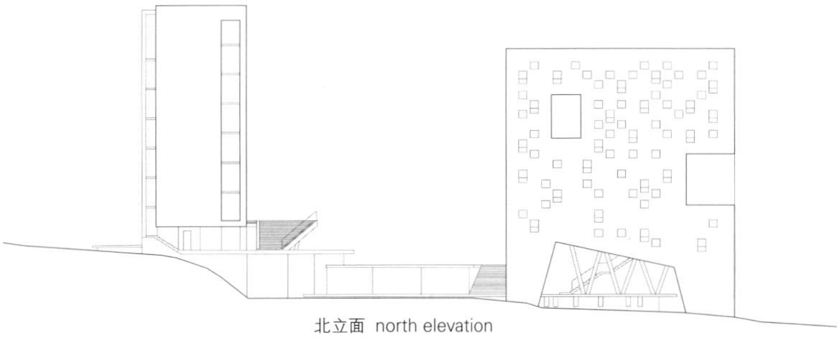

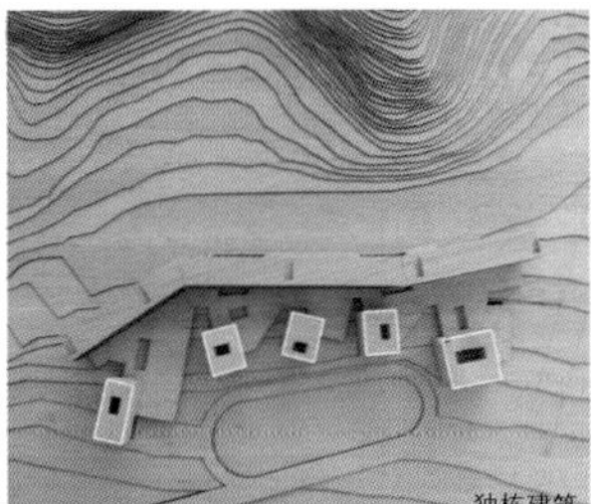

独栋建筑
single buildings

地下室
basement

公园
park

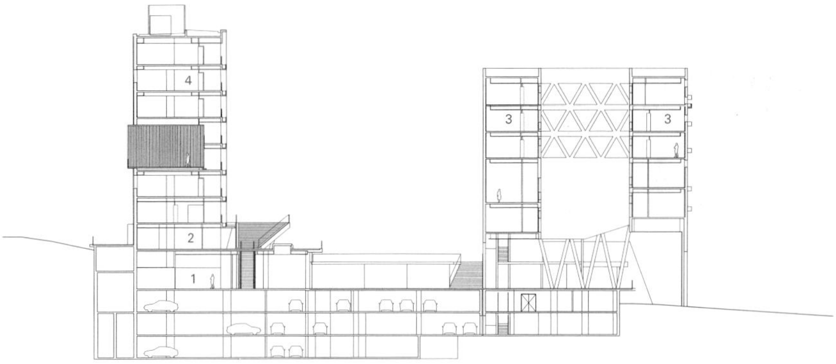

1 自习室 2 行政区 3 研究生办公室 4 教室
1. study room 2. administrative 3. graduate office 4. classroom
A–A' 剖面图 section A-A'

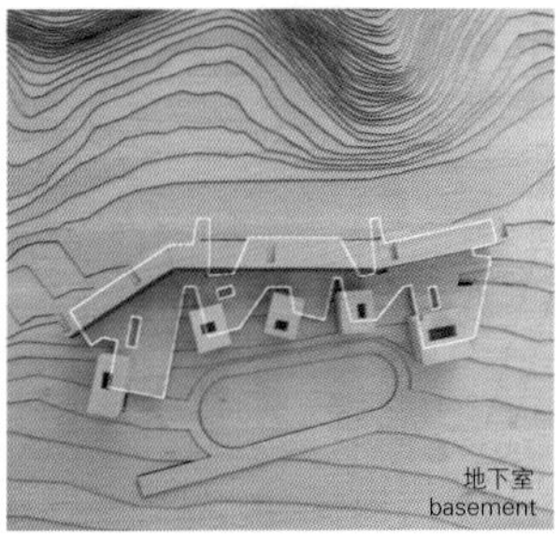

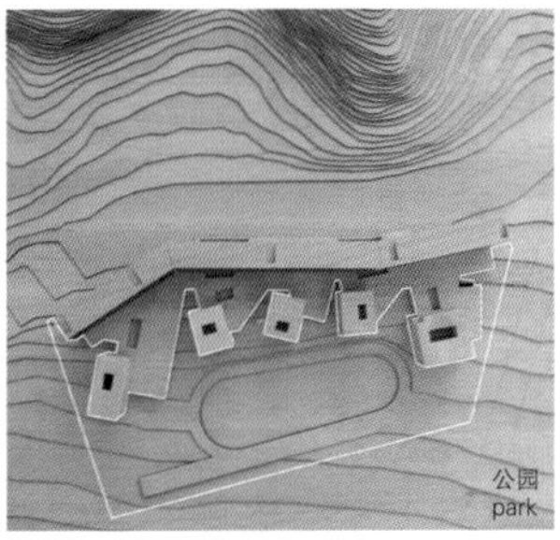

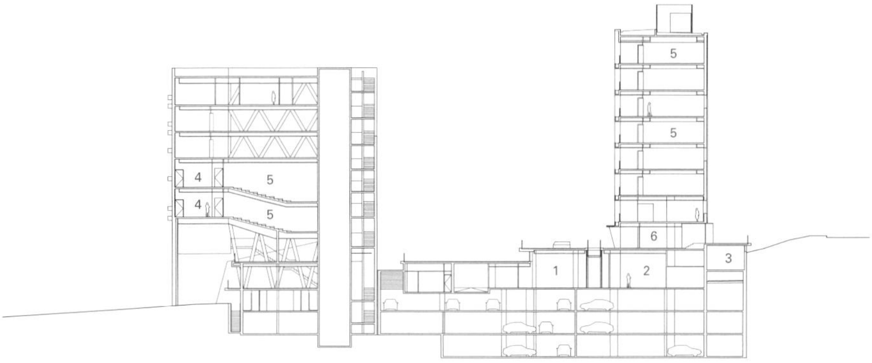

1 阅览区 2 自习室 3 厨房 4 咖啡休息室 5 教室 6 行政区
1. reading area 2. study room 3. kitchen 4. coffee break 5. classroom 6. administration
B–B' 剖面图 section B-B'

DESARROLLO
CARRERA

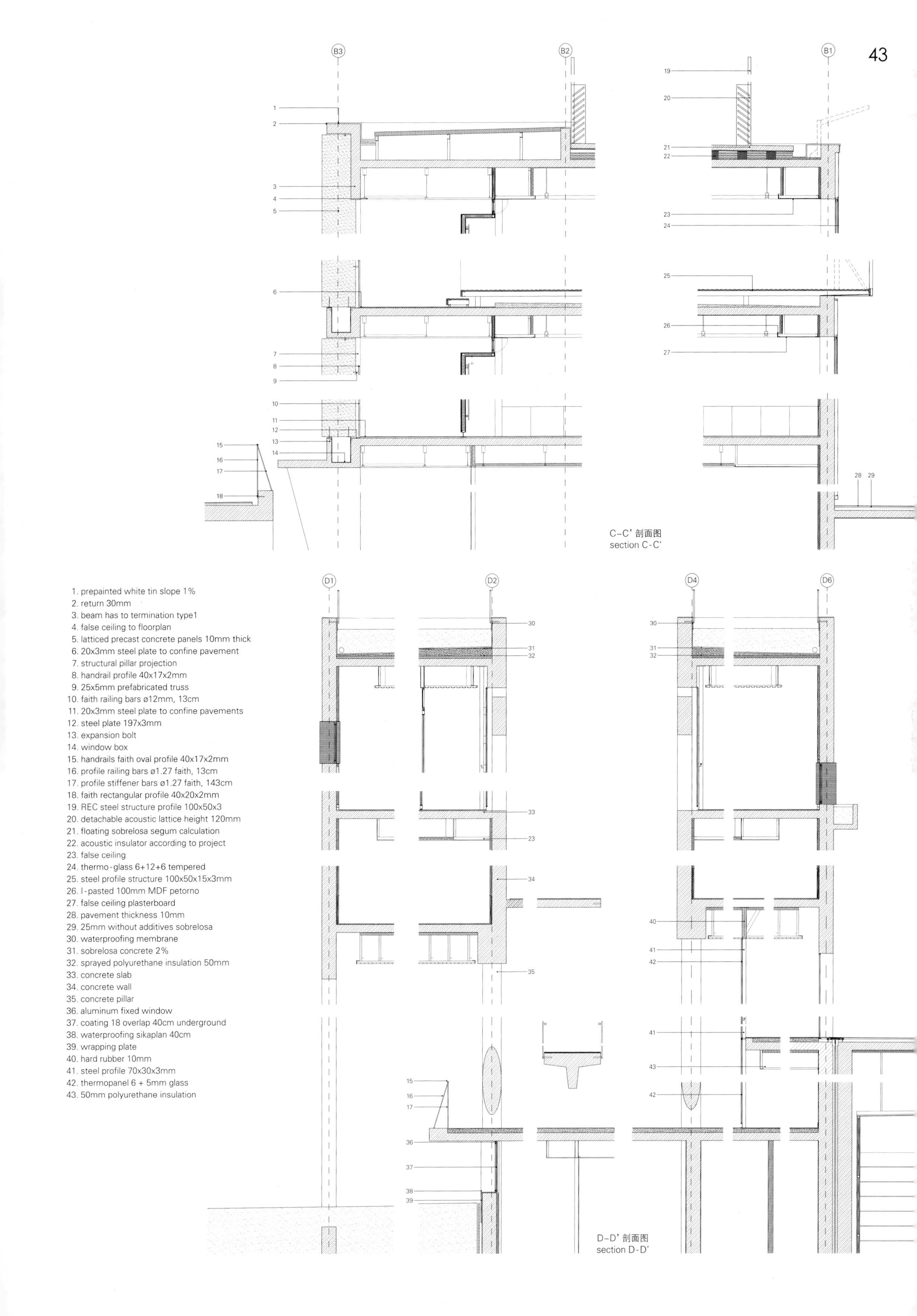
B3
B2
B1
C-C' 剖面图
section C-C'
D1
D2
D4
D6
1. prepainted white tin slope 1%
2. return 30mm
3. beam has to termination type1
4. false ceiling to floorplan
5. latticed precast concrete panels 10mm thick
6. 20x3mm steel plate to confine pavement
7. structural pillar projection
8. handrail profile 40x17x2mm
9. 25x5mm prefabricated truss
10. faith railing bars ø12mm, 13cm
11. 20x3mm steel plate to confine pavements
12. steel plate 197x3mm
13. expansion bolt
14. window box
15. handrails faith oval profile 40x17x2mm
16. profile railing bars ø1.27 faith, 13cm
17. profile stiffener bars ø1.27 faith, 143cm
18. faith rectangular profile 40x20x2mm
19. REC steel structure profile 100x50x3
20. detachable acoustic lattice height 120mm
21. floating sobrelosa segum calculation
22. acoustic insulator according to project
23. false ceiling
24. thermo-glass 6+12+6 tempered
25. steel profile structure 100x50x15x3mm
26. I-pasted 100mm MDF petorno
27. false ceiling plasterboard
28. pavement thickness 10mm
29. 25mm without additives sobrelosa
30. waterproofing membrane
31. sobrelosa concrete 2%
32. sprayed polyurethane insulation 50mm
33. concrete slab
34. concrete wall
35. concrete pillar
36. aluminum fixed window
37. coating 18 overlap 40cm underground
38. waterproofing sikaplan 40cm
39. wrapping plate
40. hard rubber 10mm
41. steel profile 70x30x3mm
42. thermopanel 6 + 5mm glass
43. 50mm polyurethane insulation
D-D' 剖面图
section D-D'

company the university in its commitment and transcendence. The density of the volumes, the concrete that lasts and ages, walls to be covered with vines showing the passing of the seasons, and a park that matures in years and stone courtyards, combine to consolidate over time. In addition, the project seeks to build a connection beyond its neighbors, with its geographic environment. It takes advantage of the slope to render the courtyards dominant over the territory, and builds terraces at different heights, as well as a roof garden, that connects the everyday life of the project with the distant geography, San Cristobal Hill, and Huechuraba Valley. These are places that link leisure time with the surrounding landscape.

The project is part of a strong desire of the university to build environmentally sustainable infrastructure. The project is designed from the architecture, landscape, and technical facilities, in terms of savings in energy and water, minimizing its impact on the environment, and insuring high standards of comfort for its occupants. The building seeks to reduce solar impact on the shell: the exterior walls are constructed with highly controlled openings, just enough for good lighting and views, and a central void for indirect light, around which the circulation is laid out. Furthermore, the building proposes a system of planters on the north and west facades, to grow deciduous vines that provide shade and humidity during the hot months, and a roof garden that acts as a buffer for thermal insulation. In the undergraduate building, the east facade is constructed from a concrete wall clad in decorated glass reflecting the hill and filtering light into the classrooms through controlled openings. To the west is a large vertical concrete lattice, with perennial vegetation planters, shading the corridor and classrooms. Finally, the building seeks to exploit the breeze from the west in the hot months, to generate cross ventilation in the classroom building and upward movement of air in the central void of the graduate building.

项目名称：School of Economics and Business Diego Portales University
地点：Santa Clara, Huechuraba, Santiago Metropolitan Region, Chile
建筑师：Rafael Hevia, Rodrigo Duque Motta, Gabriela Manzi
项目管理：Unidad Servicios Externos UDP
合作者：Catalina Ventura
结构工程师：Luis Soler P. & Asociados
景观建筑师：Francisca Saelzer
可持续性顾问：Edificioverde S.A.
照明：Monica Pérez & Asociados
施工单位：Bravo e Izquierdo
技术检测：Inspecta S.A.
用地面积：22,737m²
总建筑面积：16,644 m²
设计时间：2010
施工时间：2011
竣工时间：2013
摄影师：©FG+SG Architectural Photography

卡塔尔乔治城涉外事务学院

Legorreta + Legorreta

卡塔尔基金会教育城是一个令人振奋且雄心勃勃的项目，由卡塔尔国家元首的第二任妻子谢赫·莫扎主持成立。该项目将会成为中东以及世界其他地区未来教育项目的开山之作。教育城将汇集来自世界各地的设计师设计的各种风格的独立建筑。

日本建筑师矶崎新负责该项目的总体规划，旨在建造一处建筑与外部环境之间能够和谐交织的建筑集群。一条条自北向南、自东向西且穿过其间的"绿脊"是该项目的主要特点。

Legorreta+Legorreta建筑事务所负责设计德州农工大学工程学院（55 000m²）、卡内基梅隆商学院（40 000m²），以及乔治城涉外事务学院的计算机科学学院（3716m²）和学生中心（32 182m²）项目。庭院及公共空间的设计增强了学生及教职人员之间的互动。流线设计促进了人们环绕建筑开展活动，由此形成非正式的会客空间。传统元素，如拱形游廊、庭院和喷泉都应用到了项目之中。

埃蒙德·A.沃尔什涉外事务学院的乔治城大学卡塔尔校区整体坐落于教育城北侧。西侧相邻的建筑是中央图书馆，西南侧是学生中心，南侧有一处遗迹废墟和一座公园，它们通过建筑的景观设计与方位实现了视觉和概念上的强烈联系。

设计主旨是打破整体建筑纪念碑式的风格，而改为一种更为人性化的、使学生们感觉到舒适的，并传递"家的感觉"的良好环境氛围。建筑由各种更小的院系构成，是为了给人以村庄的感觉。建筑的所有部分都与涉外事务学院的核心公共区连接。入口借由雕塑装饰着的人行步道与南部的"绿脊"连接。

景观美化的庭院和门廊空间随处可见，装点着建筑群，意在为日常生活增添宁静之感，同时增强了在公共区域欣赏这葱郁的景观绿洲的私密感。

外部环境也给予了特别的关注。水景设计不但使空间重新恢复活力，同时还起到净化的作用。一座仙人掌花园沿北侧立面设置，成为沙漠与设计区之间的过渡标志。庭院为相邻建筑的功能增色，充当了非正式的休闲娱乐过渡区域。

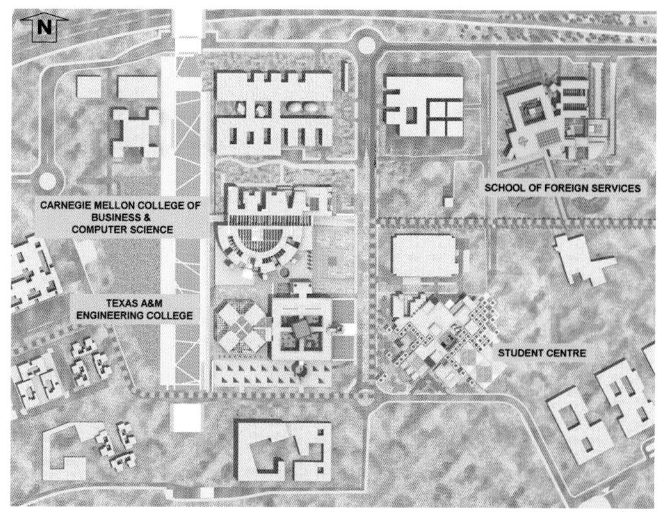

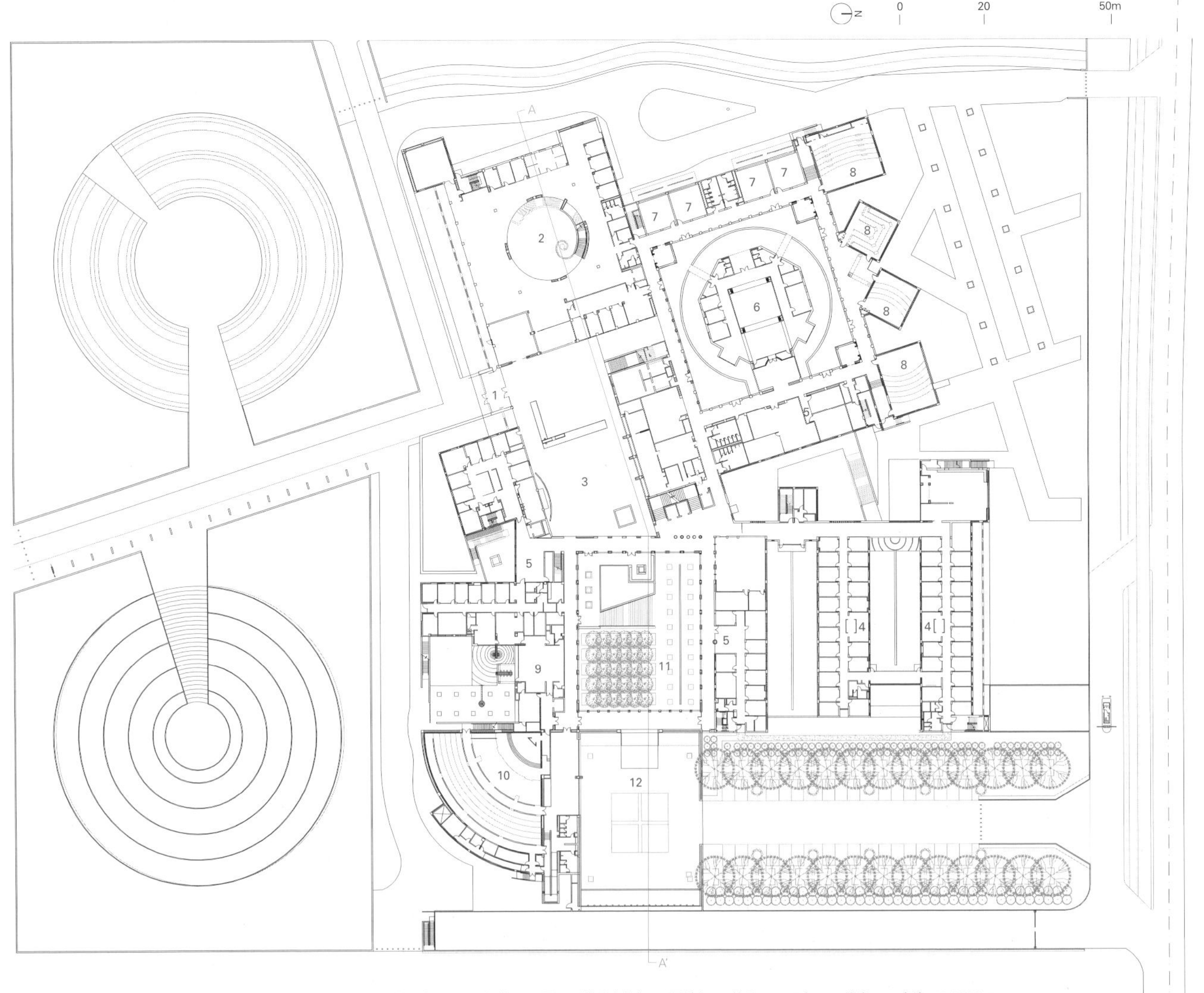

1 接待处 2 图书馆 3 公用区 4 教师办公室 5 行政区 6 国际区域研究中心 7 研讨室 8 教室 9 VIP室 10 礼堂 11 庭院 12 VIP入口
1. reception 2. library 3. common area 4. faculty offices 5. admin 6. center for International Regional Studies 7. seminar 8. classroom 9. VIP 10. auditorium 11. courtyard 12. VIP entrance

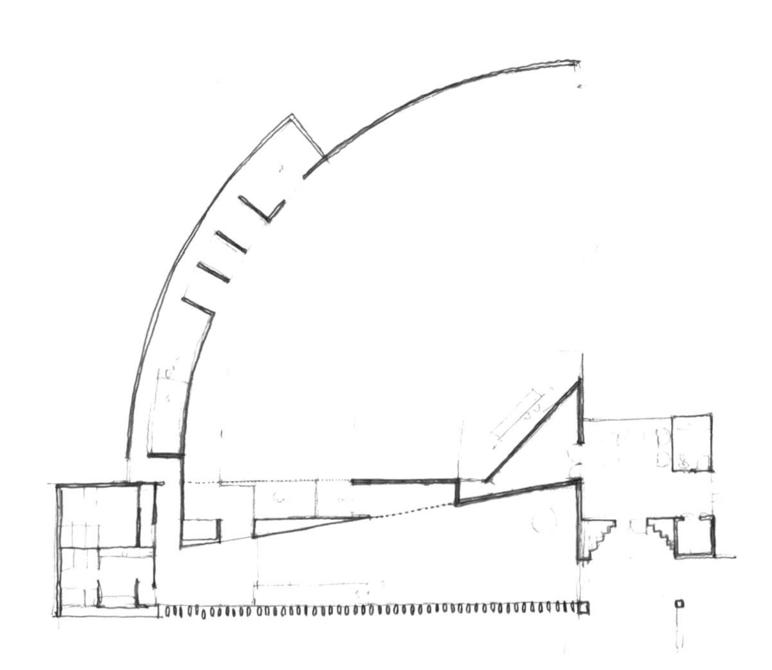

Georgetown School of Foreign Service in Qatar

The Qatar Foundation Education City is an exciting and ambitious project spearheading by Sheika Mozah, the second wife of the Emir. It will be the keystone for future educational projects in the Middle East and other parts of the world. The Education City will be composed of various unique buildings designed by architects selected from around the world.

The Master plan, designed by Arata Isosaki, is a harmonious blend of architecture and exterior spaces. The main features are the Green Spines crossing from north to south as well as from east to west.

Legorreta+Legorreta was selected to design the Texas A&M Engineering College (592,000 sq ft), the Carnegie Mellon Business (430,550 sq ft) and Computer Science College, the Georgetown School of Foreign Service (40,000 sq ft) as well as the Student Center (346,404.17 sq ft). The design of the courtyards and public spaces enhances the interaction between students and faculty staff. Circulations stimulate movement around the building and informal meeting spaces are created. Traditional concepts

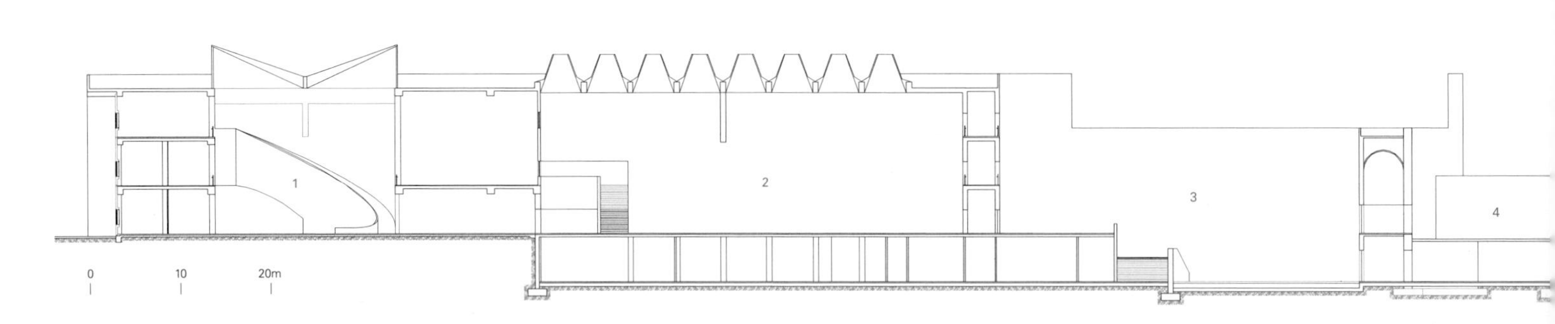

1 图书馆 2 公用区 3 庭院 4 VIP入口 5 停车场
1. library 2. common area 3. courtyard 4. VIP entrance 5. parking
A–A' 剖面图 section A-A'

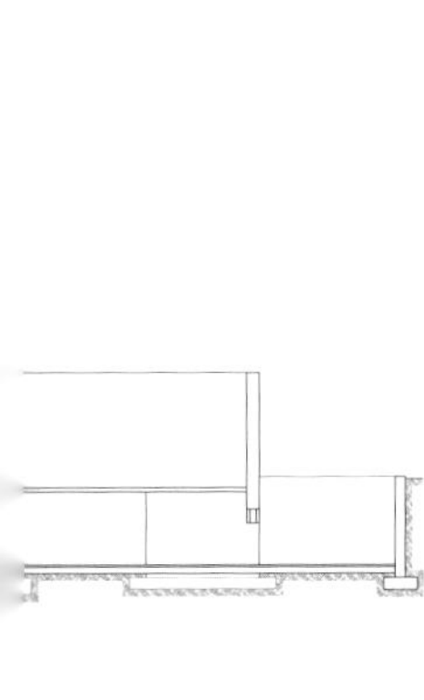

项目名称：Georgetown School of Foreign Service
地点：Doha, Qatar
建筑师：Legorreta + Legorreta
项目建筑师：Asad M. Khan _ Langdon Wilson International / 合作建筑师：Francisco Cortina
景观建筑师：Carter Romanek Landscape Architects Inc.
用地面积：49,982m² / 有效楼层面积：68,574m² / 设计时间：2006.7—2008.6 / 施工时间：2008.6—2011.1
摄影师：©Yona Schley(courtesy of the architect)

of architecture such as arcades, courtyards and fountains are integrated in the project.

The Georgetown University's Campus in Qatar of the Edmund A. Walsh School of Foreign Service is integrated within the northern side of the Education City Campus. The adjacent building to the west is the Central Library, to the southwest is the Student Center, to the south lies the Heritage Ruin and a park with a strong visual and conceptual connection achieved by landscape design and orientation of the building.

The major design intent is to break down the monumental style of the overall building to a more human scale one to make the student feel comfortable and transmit a "feel like home" atmosphere. The building is a composition of various smaller departments in order to give it a village-like character. All parts of the building are connected to the Common Space which is the core of School of Foreign Service. The entrance is connected by pleasant pedestrian walk with sculptures to the Green Spine on the South.

Landscaped courtyards and atria spaces interspersed throughout the complex of the building are intended to bring a tranquility feeling to the day to day activities and promote a sense of intimacy within the spaces that are orientated towards look onto these richly landscaped oases.

Special attention is given to the outdoor environment. Water features refresh and ventilate the spaces around them. A cactus garden is located along the north facade and marks the transition between the desert and designed areas. Courtyards compliment adjacent building's functions and are used for informal recreational and pleasant transition zones.

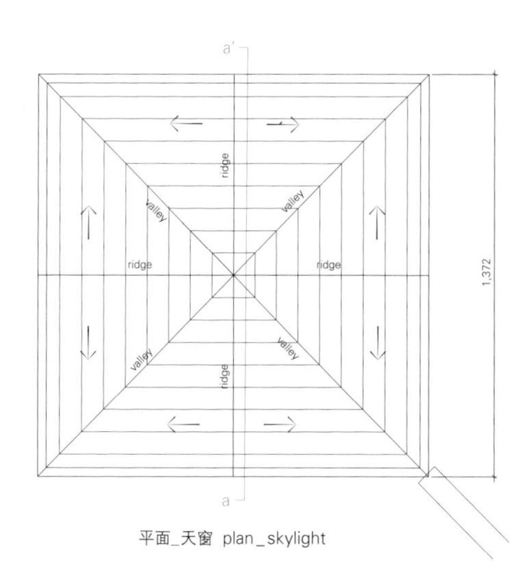

平面_天窗 plan_skylight

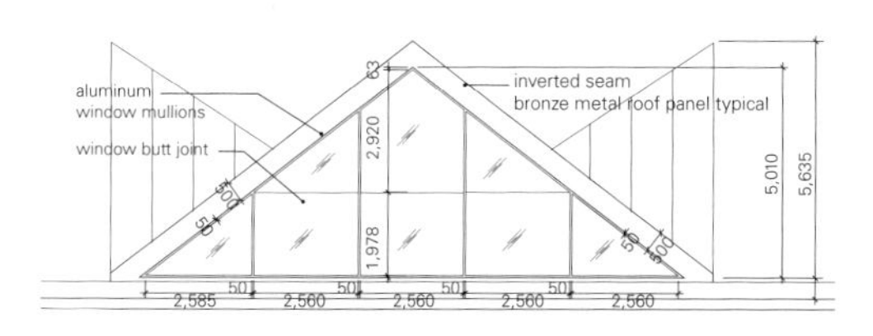

立面_天窗 elevation_skylight

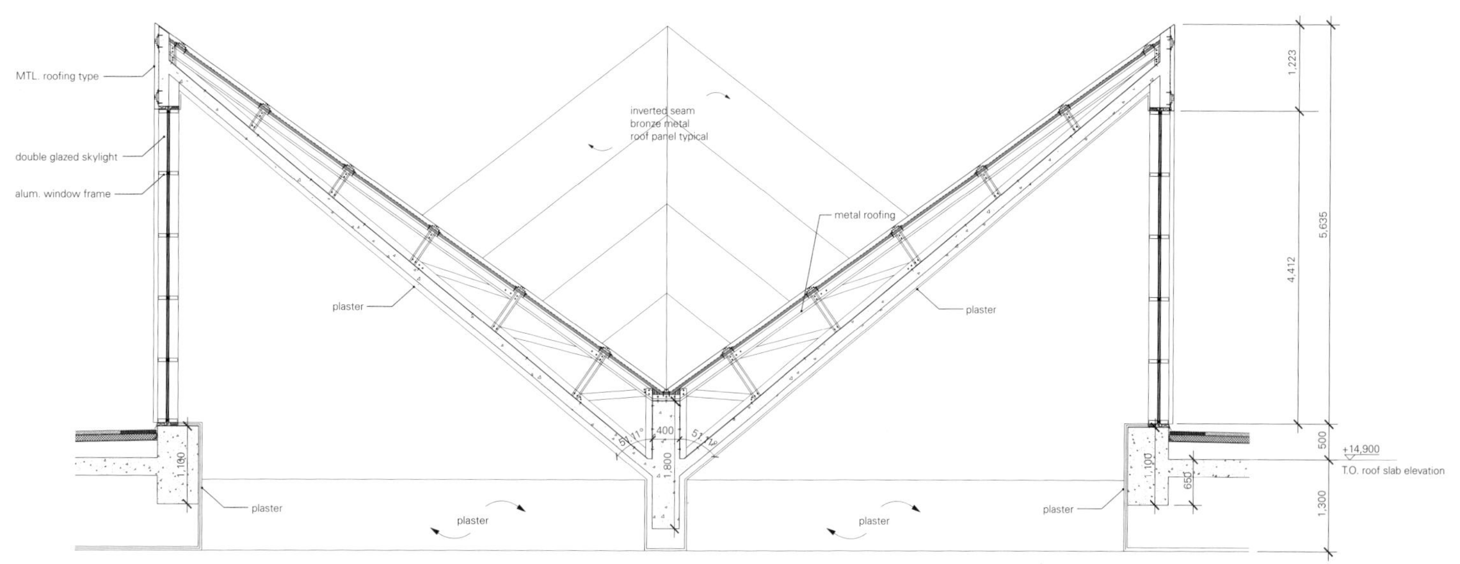

a-a' 剖面图 section a-a'

维也纳金融大学校区的教学中心

BUSarchitektur

网络

建筑师将这座位于维也纳金融大学新校园东北部的新学术空间设想为一个多元网络的汇聚点。教学中心不仅仅是一座建筑，更是一个学习平台，促进自由交流，同时依然能够承担内部进行的班级授课和活动。这种对于自由交流氛围的追求，不论其层次与规模，目的均是相同的，即修行之旅本身即是目的地。对于建筑师来说，一座现代化的教学中心应该能够在日复一日的常规学习过程中激起教授者与学习者的好奇心，求知者在这里漫步，如同登上了一座舞台，不受空间限制，给予想象力以无限畅游空间。

建筑之上：漫步

教学中心主入口前方有一个论坛广场，虽然处于校园的外部区域，但它更像是一处延伸出来的内部空间。空间的多种用途能够在教学中心内部的交流空间体会到，也能够通过教室内的阶梯和坡道得以体验。

建筑之下：流通

停车场位于中央礼堂和主讲堂之下，两座建筑均是悬浮结构设计，不受大城市熙攘喧嚣的干扰，使用者因此得以享受静心"聆听"的愉悦。静谧穿梭于运动之中。

建筑之间：探索

封闭及开放空间顺次排列，致力在这座为4000多名使用者设计的公共建筑中创造私人区域。

建筑边沿：想象

白天上课时间，会议厅通过全方位射入的日光提供照明，形成一处能够培养创造力的生机勃勃的空间。人工LED照明的使用则是利用现代科技使知识得以更广泛的传播。

建筑周围：发现

通往教室的走道是一段上升的螺旋通道，沿路有一处学习空间的绿洲。不同的学习强度吸引着人们的目光徘徊于探索来自不同方向的未知之境。

建筑之后：聆听

维也纳的咖啡文化在各类餐饮区都是360度全景可见、可闻、可观、可享、可感的，设置在各个广场之间的人流中。平日里到餐馆小饮一杯咖啡，来见证四季交替变迁，欣赏日光夜景交错，提升整体体验。

超越教学中心：全球化

建筑师设想建造一处不同层次上的致密化学习空间，或通过水平上的空间排列（露天广场–长椅–城市屋顶平台–美食广场），或通过空间的垂直分层排列（通往各个院系的通道楼梯–走廊–中庭）。全球商务

cement tiles
low growing grass
high growing grass
ornamental greenery
dense vegetation
grass covered cobblestone
ground drainage
granulated rubber
grass
gravel
cobblestone
asphalt

mensa

和贸易学院介于图书馆、学习中心和教学中心之间，有效地成为了一切活动的中心。真正的中心就是这里。

通过简单的语言和复杂的空间排列，维也纳金融大学新校园的教学中心项目实现了通过建筑风格来鼓励积极的社交活动这一构想。在这里，休闲游憩的空间和方便进行各类活动、事项和空间互动的广场共存；或者，换句话说，整体设计构想即是人、活动和建筑之间紧密联系。

Teaching Center in WU Campus

Networking

The architects envisage the new academic space located at the north-east of the new campus of the Vienna University of Economics and Business as a venue for multiple networking. More than a mere building, the teaching center is a learning platform that promotes spontaneous communication while also making it possible to undertake the classes and activities that occur within. This quest for spontaneous exchange has different levels and scales sharing one common denominator: the path is the goal. For the architects, a contemporary teaching center should encourage curiosity in the everyday routine of those who teach as well as those who learn, becoming a place where moving about and getting on stage do not lead to experience spatial boundaries by instead giving free reign to the imagination.

Above it: strolling

The forum plaza located in front of the teaching center's main entrance is as much an exterior world of the campus, as an extended interior world. Spatial versatility is experienced across "the space in-between for communication" of the teaching center's dense body, as well as via the interior staircases and ramps inside the aula.

Below it: mobilizing

The parking access is beneath the central auditoriums and the main lecture hall(audimax), which are both suspended bodies, in which the big city's noise is no longer perceived, thereby allowing users the experience "listening". Serenity is in the flow of movement.

Between it: searching

The sequence of closed and open spaces creates private areas in this public building designed for about 4,000 users.

Alongside it: imagining

During the course of the day, the assembly hall is transformed by natural light entering from all directions, forming a lively space

that fosters creativity. The provision of artificial LED lighting enables knowledge to spread further through contemporary technology.

Around it: discovering

The walking path to aula follows an ascending spiral course accompanied by an oasis of study spaces. Different learning intensities allow people's gaze to wander the wonderland absorbed by searches in different directions.

Behind it: listening

Vienna's coffee culture can be seen, smelled, watched, enjoyed and experienced in the 360° panoramic views from the multitudinous dining area, located in between the flows from the different plazas. The daily visit to the canteen is stimulated by watching the passing of the four seasons, as well as different daytime and night scenes, which enhance the overall experience.

Beyond the Teaching Center: globalizing

The architects propose a densification of learning at different levels, either through horizontal spatial sequences (patio plaza–lounge–roofed urban platform–food court), or through a vertical spatial stratification (access staircase to the departments building–corridors–atrium). The Department of Global Business and Trade effectively becomes the center of all activities in between the library, learning center and the teaching center. The center is here.

Through a simple language and complex spatial sequences, the Teaching Center at Vienna's new WU Campus makes it possible to recognize the signs that encourage society to form by means of architecture. Here, places for leisure and recreation appear together with plazas that allow interactions between movements, events and spaces; or, in other words, connections among people, activities and architecture are conceived. BUSarchitektur

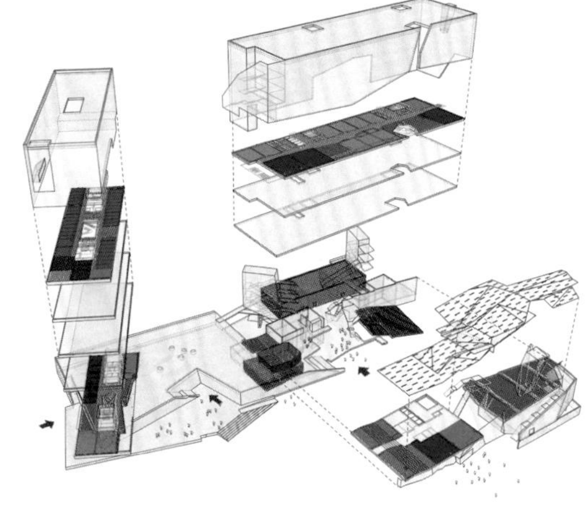

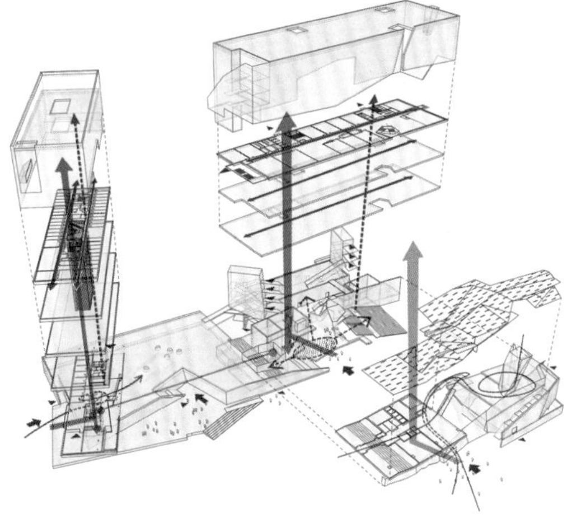

项目名称：Teaching Center in WU Campus
地点：Welthandelplatz 1-1020, Vienna
建筑师：BUSarchitektur
甲方：Projektentwicklungsgesellschaft Wirtschaftsuniversität Wien Neu GmbH
获得奖项：1st Prize in the International Competition for the Masterplan and Executive Project for WU Campus
可用面积：28,349m² / 建筑面积：32,484m² / 体积：149,063m³
造价：EUR 52.4 M / 设计时间：2008 / 竣工时间：2013
摄影师：©BOAnet(courtesy of the architect)

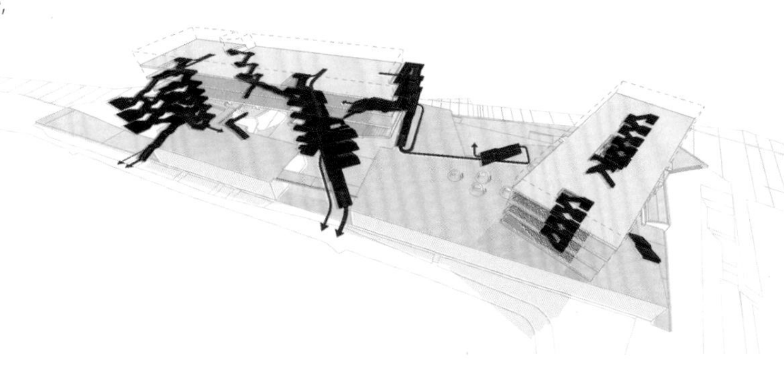

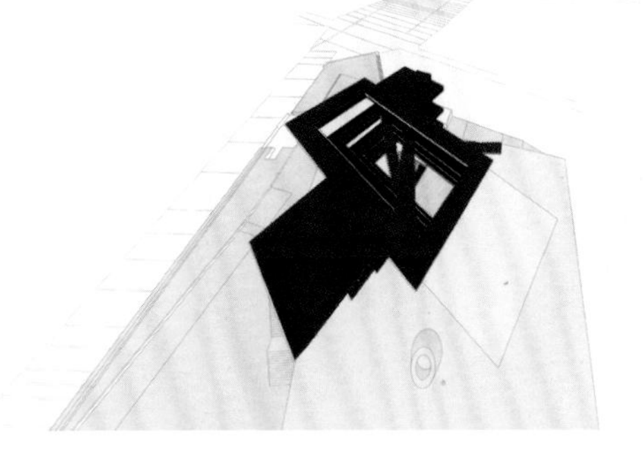

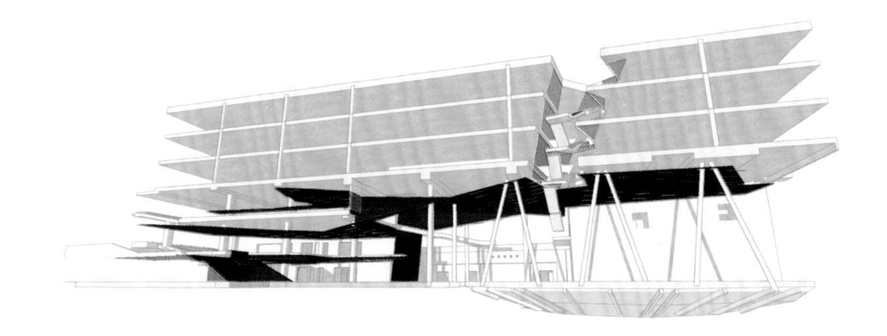

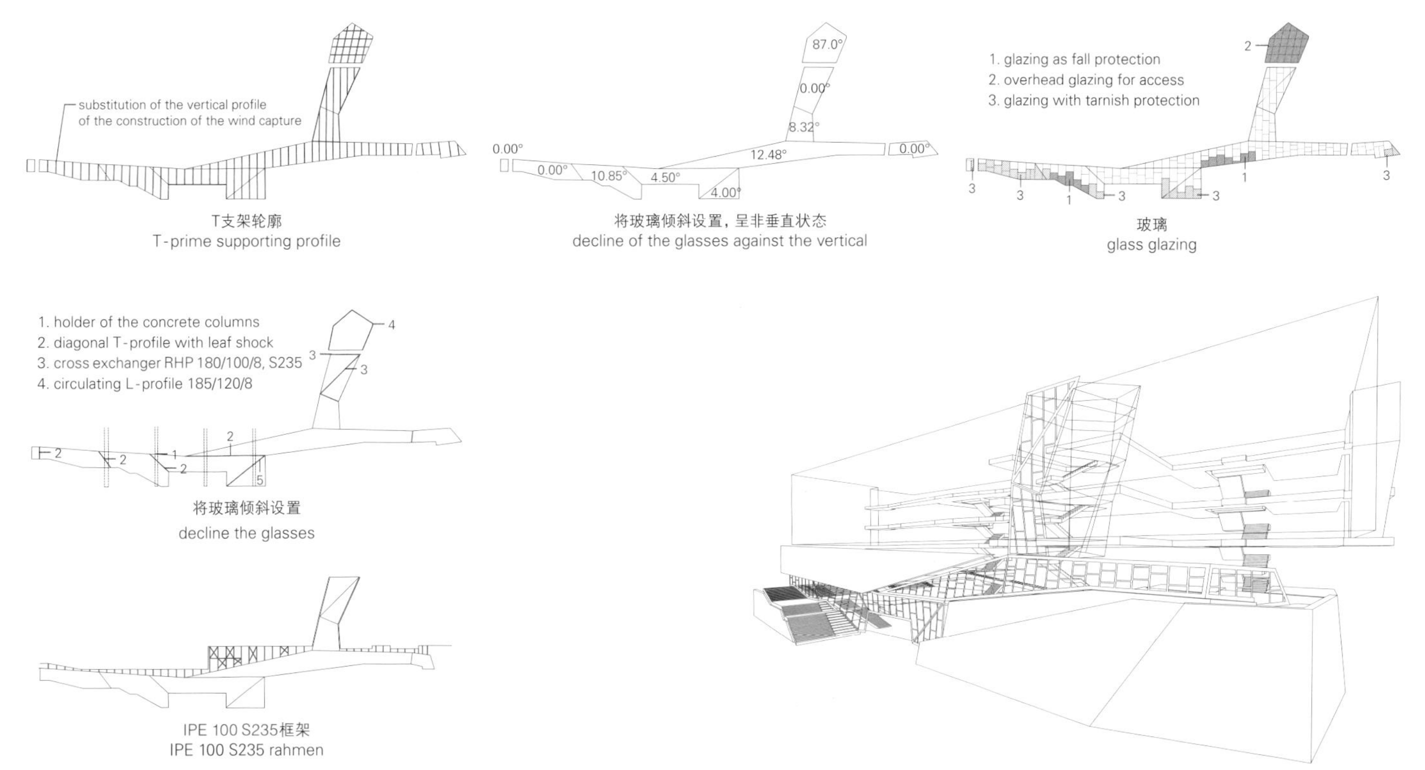

T支架轮廓
T-prime supporting profile

将玻璃倾斜设置，呈非垂直状态
decline of the glasses against the vertical

玻璃
glass glazing

将玻璃倾斜设置
decline the glasses

IPE 100 S235框架
IPE 100 S235 rahmen

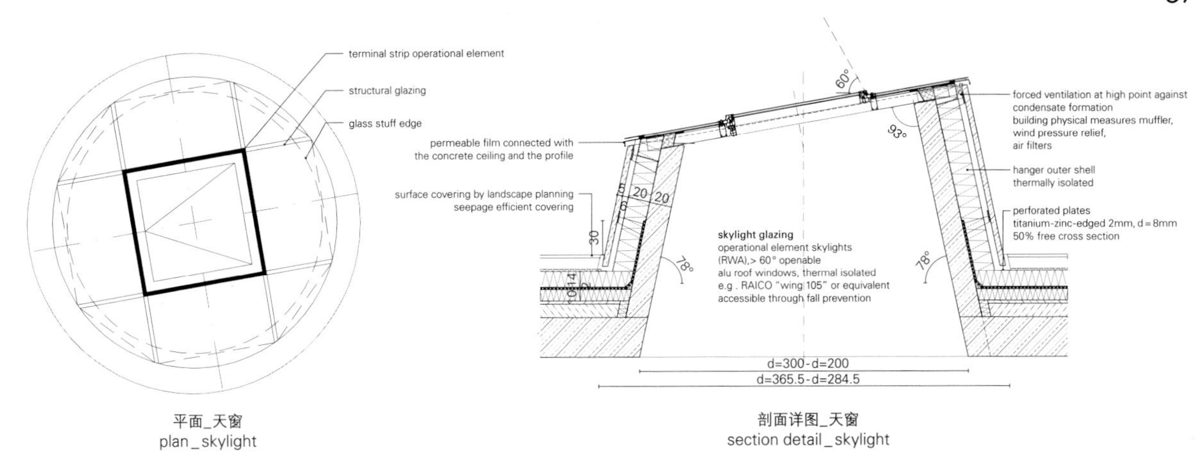

平面_天窗
plan_skylight

剖面详图_天窗
section detail_skylight

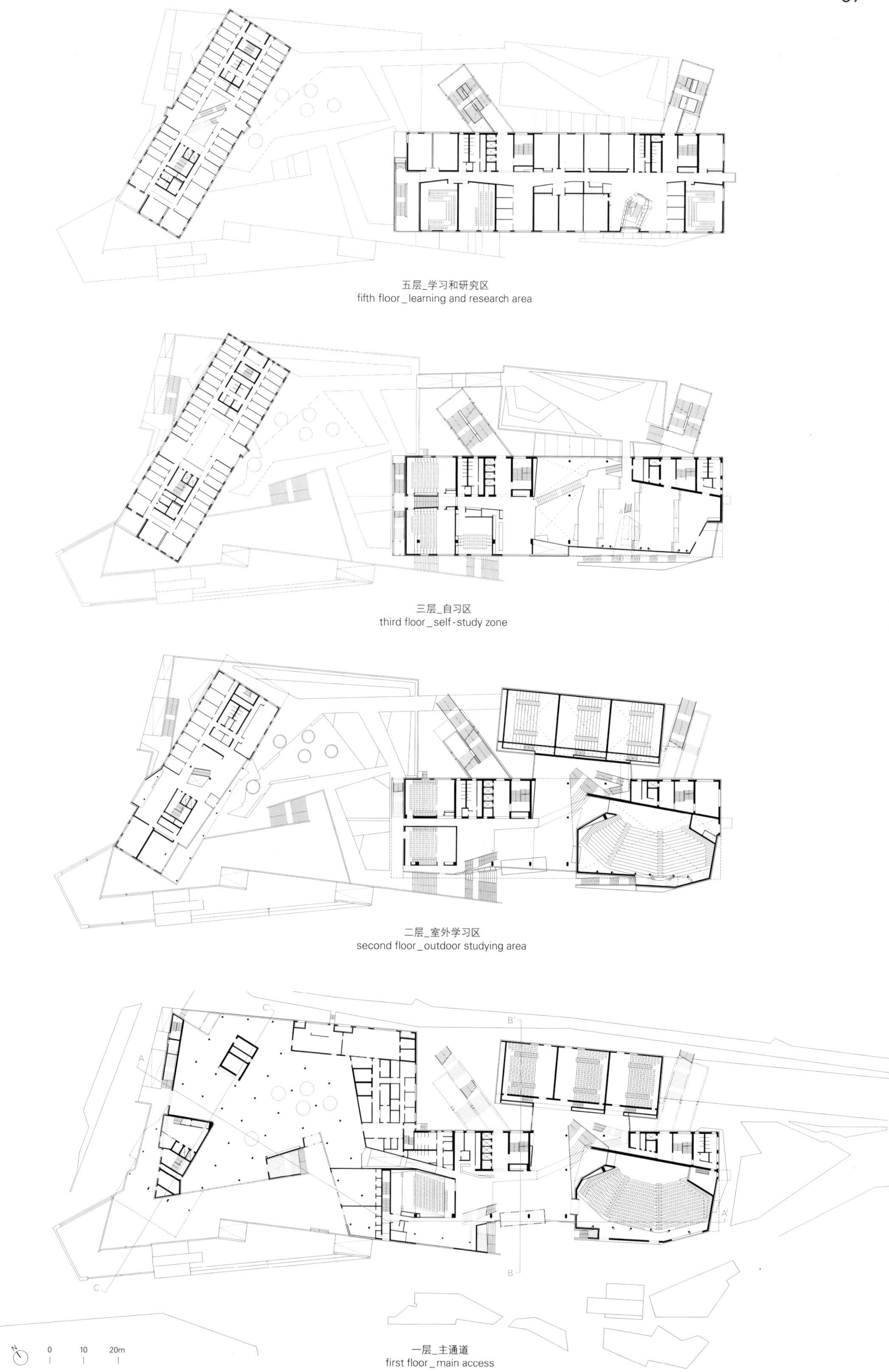

五层_学习和研究区
fifth floor_learning and research area

三层_自习区
third floor_self-study zone

二层_室外学习区
second floor_outdoor studying area

一层_主通道
first floor_main access

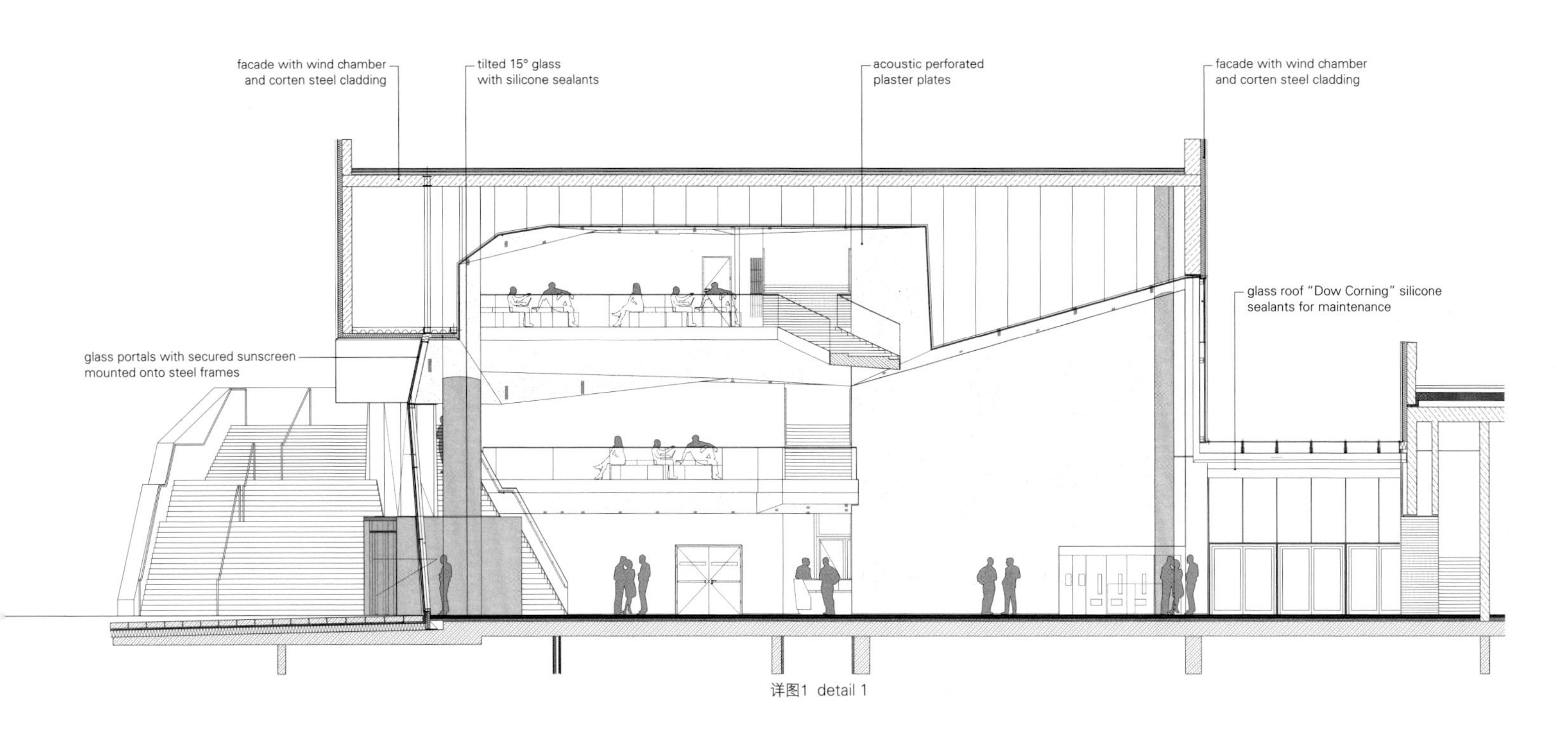
facade with wind chamber
and corten steel cladding
tilted 15° glass
with silicone sealants
acoustic perforated
plaster plates
facade with wind chamber
and corten steel cladding
glass roof "Dow Corning" silicone
sealants for maintenance
glass portals with secured sunscreen
mounted onto steel frames

详图1 detail 1

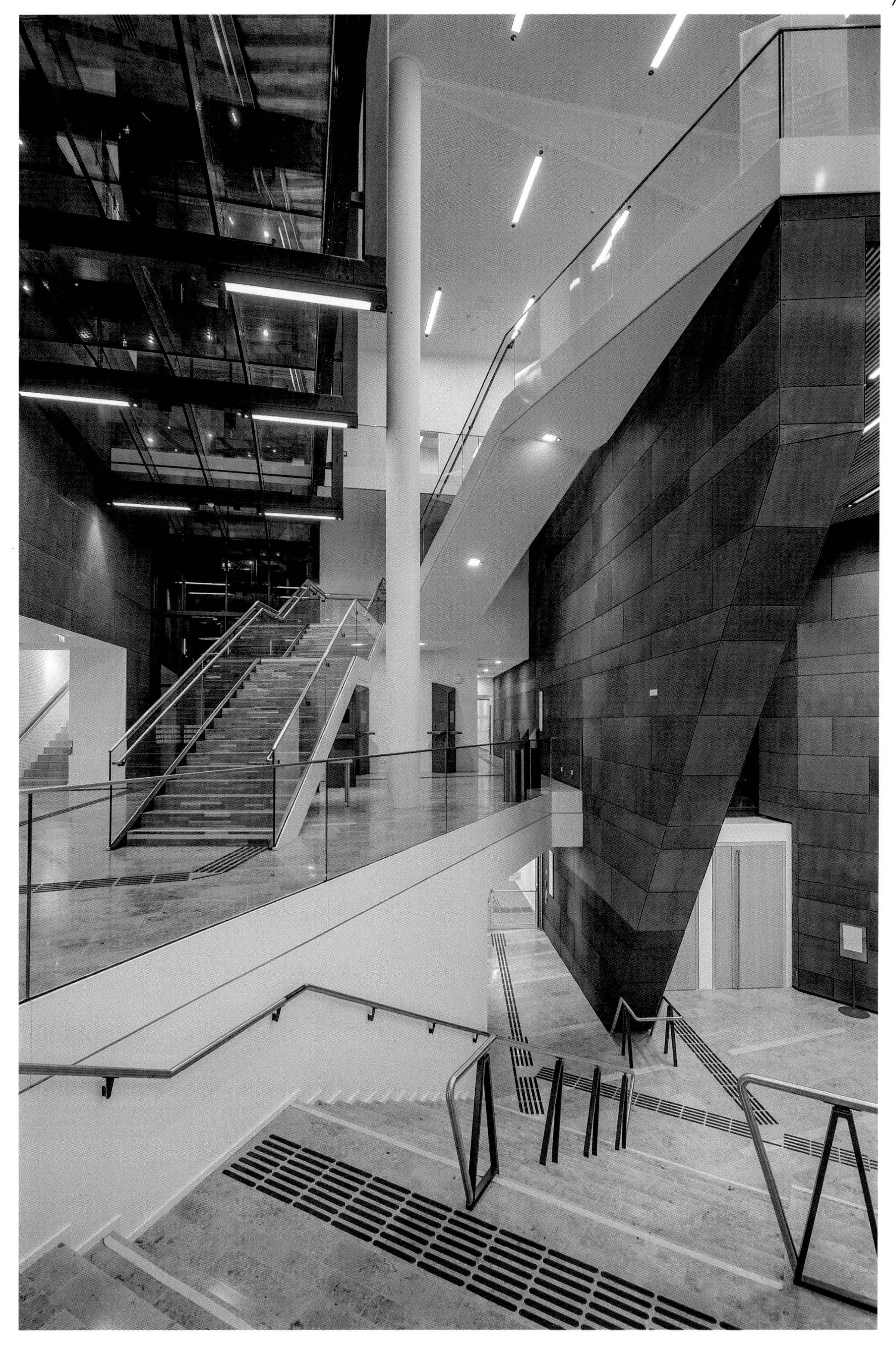

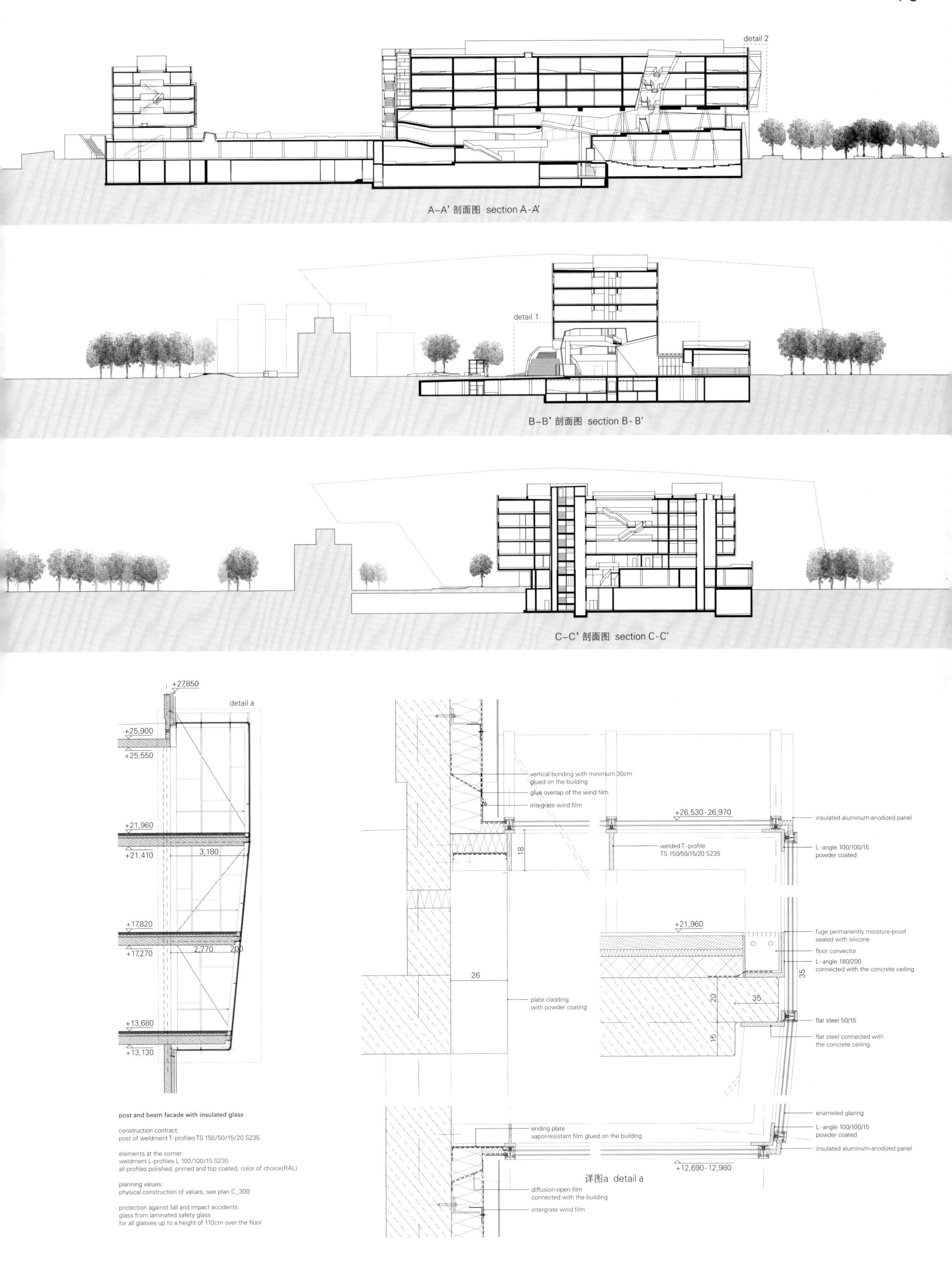

详图a detail a

post and beam facade with insulated glass

construction contract:
post of weldment T-profiles TS 150/50/15/20 S235

elements at the corner:
weldment L-profiles L 100/100/15 S235
all profiles polished, primed and top coated, color of choice(RAL)

planning values:
physical construction of values, see plan C_300

protection against fall and impact accidents:
glass from laminated safety glass
for all glasses up to a height of 110cm over the floor

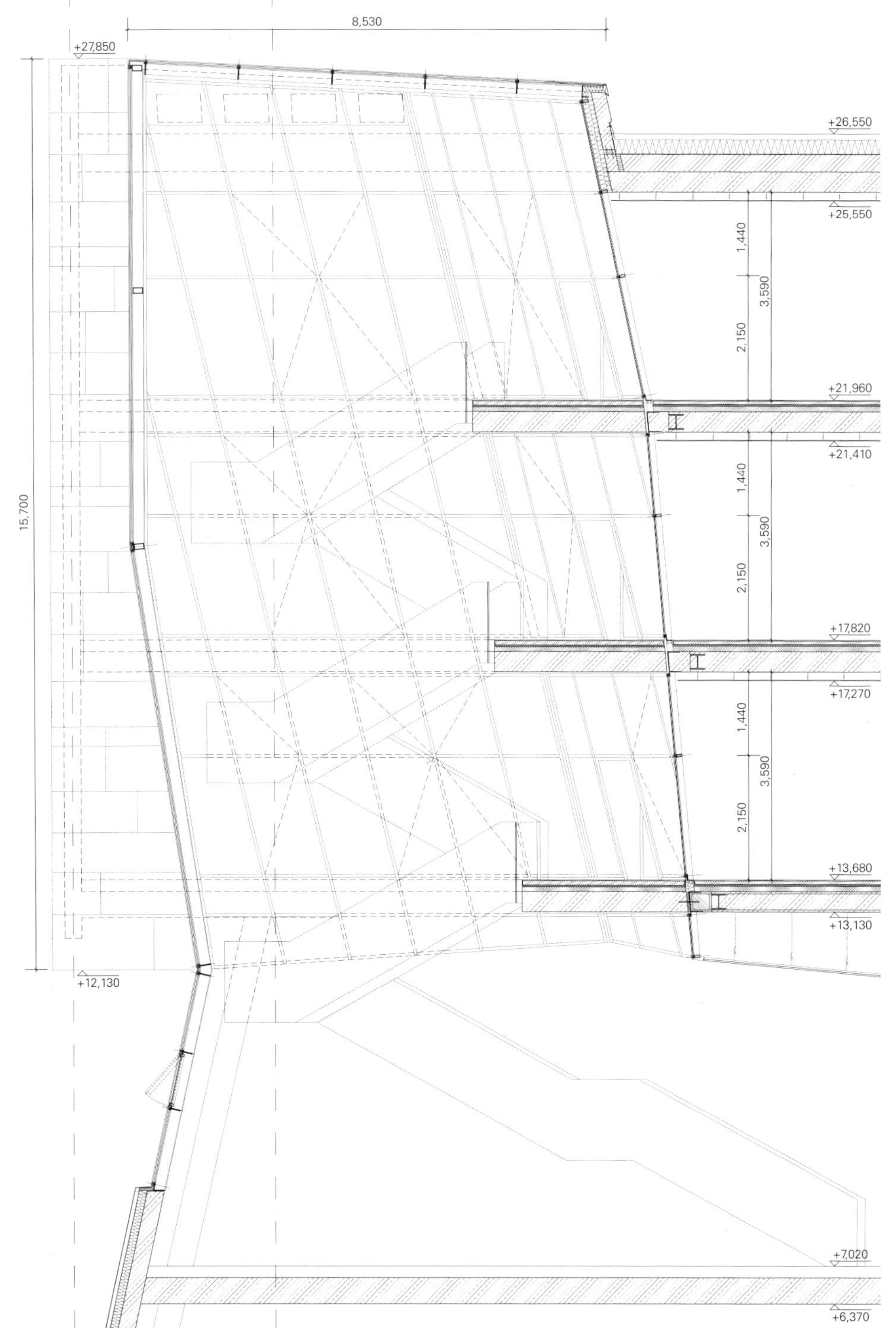

a–a' 剖面图 section a-a'

post latch inside facade staircase

construction contract:
vertical uprights full steel 50x100, S235;
floor, 1 transitory tie RHP 50/100/5,
weldment S235
all profiles plained, fire protection painting,
color of choice (RAL)

joined by tension rod D = 8mm steel,
fire protection painting with color of choice (RAL)

glass holder by post-and-beam system
e.g . RAICO Therm S-I or equivalent

glazing:
single glazed, held on 4 sides
in the area of the floor structure's mirrored glass;
glass panels in the overhead area (> 15 °)
and glass panels with requirement of fall protection
see det. NO . C302_02

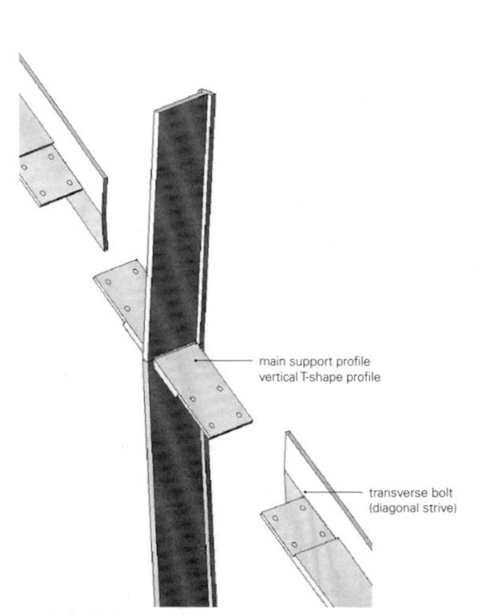

节点轴测图 nodal point axonometric

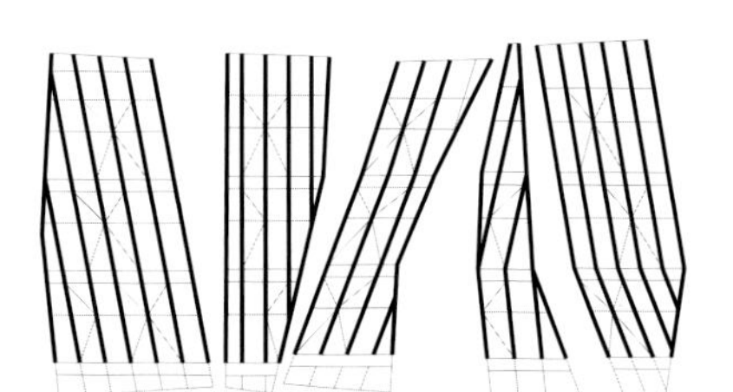

垂直全钢 50×100
vertical uprights full steel 50×100

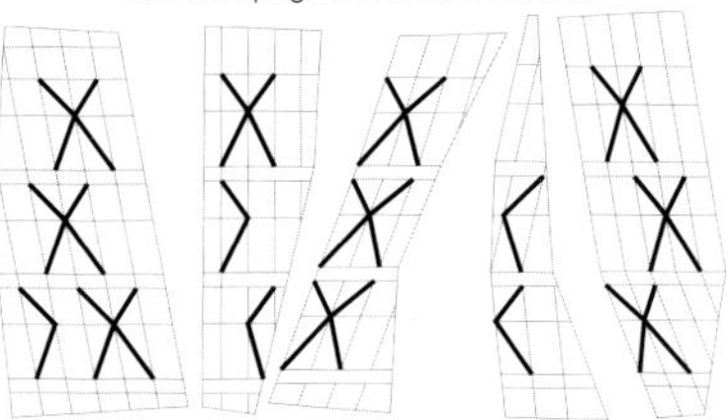

加入受拉构件 d=8mm
joined with tension member d = 8 mm

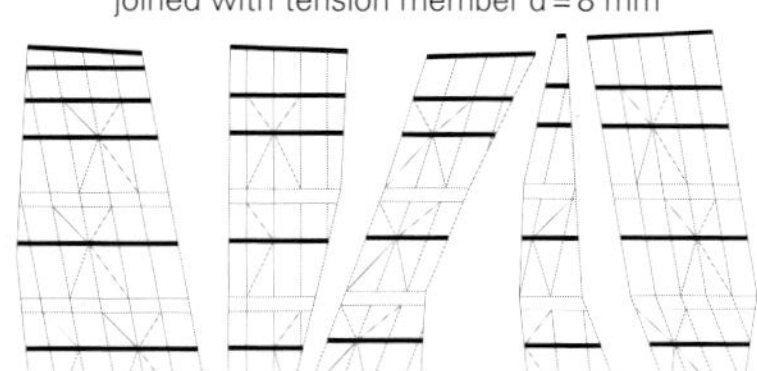

托架 RHP 50/100/5
carrier RHP 50/100/5

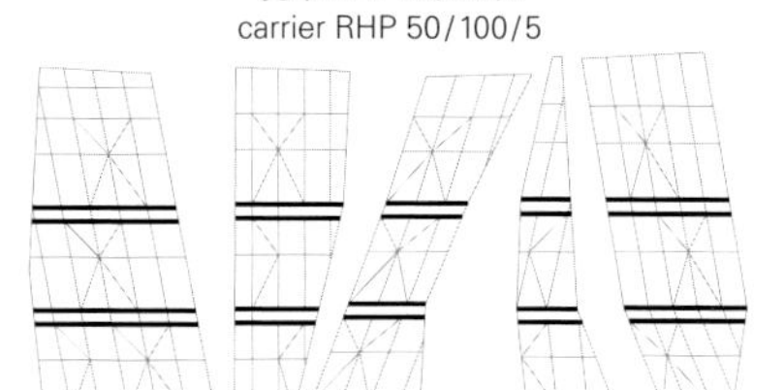

型材，位于地下室天花板一端
profiles as an ending for basement ceiling

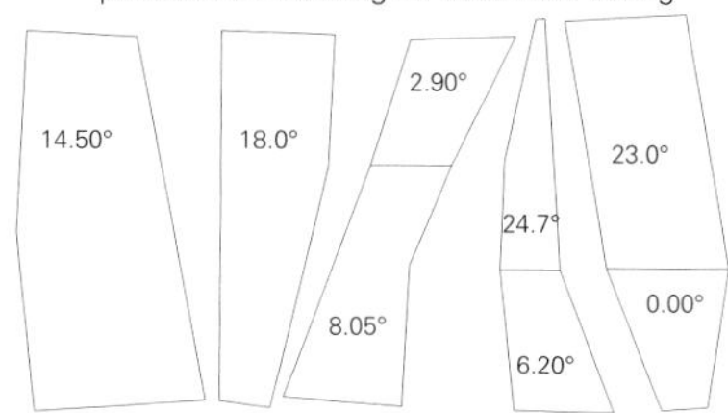

将玻璃倾斜设置，呈非垂直状态
decline the glasses against the vertical

防火玻璃
fire protection glazing

层压安全玻璃，位于头顶处（大于15°）
laminated safety glass, over head glazing (>15°)

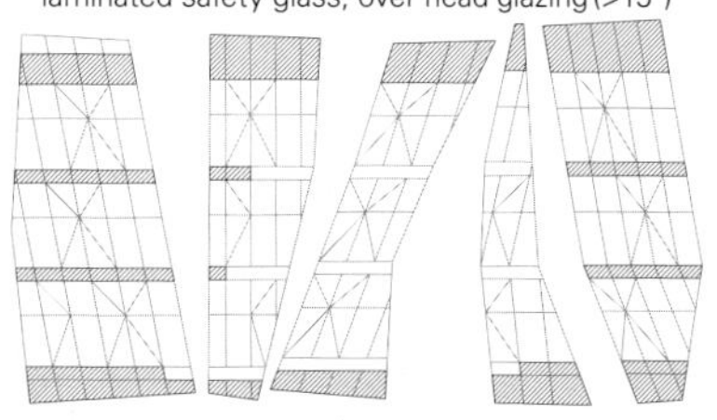

镜像玻璃
mirrored glazing

伦敦艺术大学中央圣马丁艺术与设计学院

Stanton Williams Architects

照片提供：©John Sturrock(courtesy of the architect)

中央圣马丁学院的总院伦敦艺术大学位于271 139m²的场地内，它第一次将学院的所有活动整合在一个屋顶之下，且其总体规划迅速将英王十字车站改造成伦敦的一处全新街区。

中央圣马丁学院增建的一座牢固的新建筑，连接其南端的二级列管谷仓，这座破旧的谷仓是该地区工业化历史遗留下来的产物。学院为坚固的六层立方体建筑，其两侧为两座平行设置的、180m长的货栈，货栈曾被用作铁路、船舶及马匹运输粮食的交通和中转站。

新校园的设计从其维多利亚时期建筑背景的雄心和规模中汲取灵感，引进了一个有力的现代化嵌入结构，使其在细节上刻意凸显新旧并置的设计特征。

谷仓建筑本身进行了修复并成为学院的主门面，其功能是作为学校的图书馆和画廊，后面的货栈则被改造成为工作间。新增建的两座主要的四层工作室建筑占据了两个货栈之间的空间，位于中央大街的两侧，覆以半透明的ETFE材质的屋顶。

根据设计构想，街道充当了建筑的中央流线的"脊柱"及富有创意的"心脏"，成为学生生活的舞台，用于日常展览和表演。桥梁连接两侧的各个核心区和工作区，为人们提供了会议、观看和思想交流的室外区域。

北端的平台剧院综合设施包括表演空间、剧院、舞蹈工作室、休闲吧和售票处。另一条覆顶的街道位于新建筑的南端，且与北端谷仓建筑平行。运转的电梯使人联想到了谷物的垂直移动，实现了这座综合设施的最初目的。地板的细节设计或保留了原有的转盘，或暗示了它们的历史位置。

这些空间和阶梯教室、学习区、餐厅、展览区以及一个屋顶平台都十分坚固耐用，处处实现了自然采光，且适应性强，原生态。这是一个变革的舞台，一个可以由教工和学生们灵活策划的空间框架。在这里，新型互动和嵌入的结构、机遇和试验能够在规范中迸发出丰富学生体验的推动力。

UAL Campus for Central Saint Martins College of Arts and Design

The University of the Arts London campus, home of Central Saint Martins College, unites the college's activities under one roof for the first time within the context of a 67 acre masterplan that is rapidly transforming King's Cross Station into a new urban quarter for London.

It provides Central Saint Martins with a substantial new building, connected at its southern end to the Grade II listed Granary Building, a rugged survivor of the area's industrial past. This solid, six-storey cubic mass is flanked by two parallel 180-metre long Transit Sheds, once used for the transport and transfer of grain by rail, boat and horse carriage.

Drawing inspiration from the ambition and scale of its Victorian architectural setting, the design of the new campus introduces a

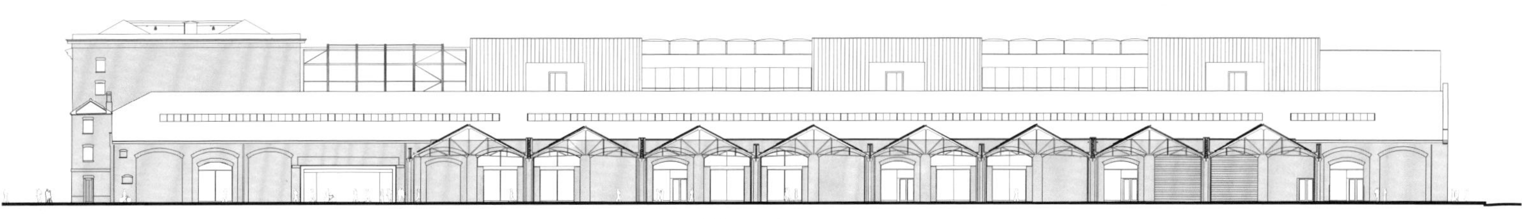

东立面 east elevation

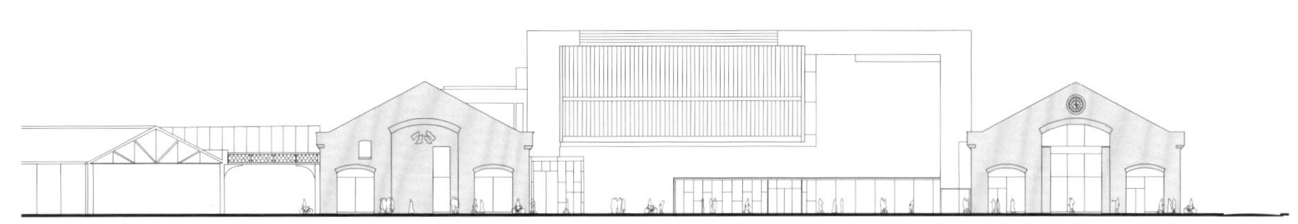

北立面 north elevation

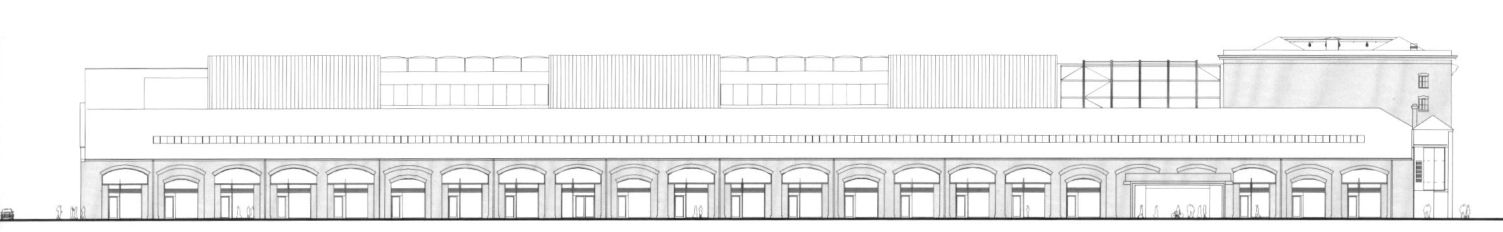

西立面 west elevation

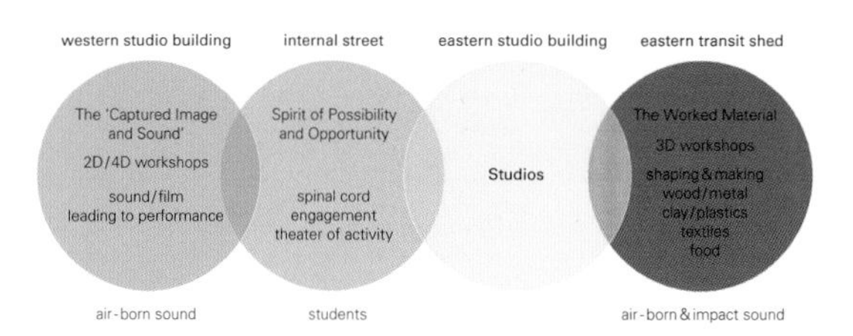

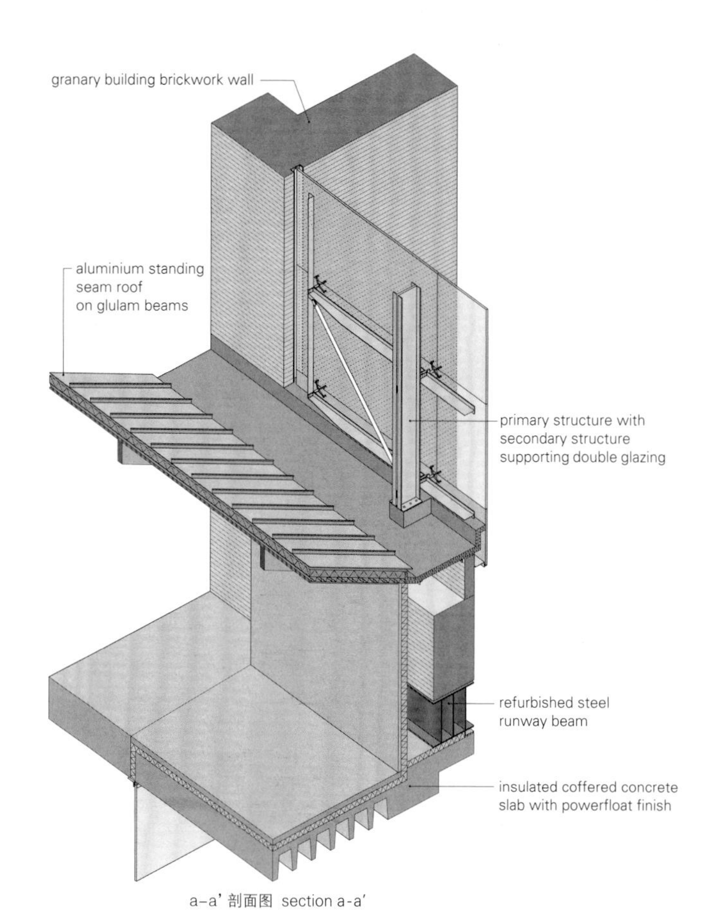

a–a' 剖面图 section a-a'

strong contemporary intervention that celebrates the juxtaposition of old and new in its detailing.

The Granary Building itself has been restored as the main front of the college, adapted to function as the college's library and galleries, while the transit sheds behind have been converted into workshops. A major addition of two new four storey studio buildings occupies the space between the two transit sheds, arranged at either side of central street, covered by a translucent ETFE roof. Conceived as an arena for student life, the street acts as a central circulation spine and creative heart for the building, used for exhibitions, shows and performances. Bridges link the various cores and workspaces, offering break-out areas for meeting, people-watching and exchanging ideas.

At the northern end, the Platform Theatre complex houses performance spaces, theatre, dance studios, foyer bar and ticket office. At the southern end of the new block and running parallel with the north end of the Granary Building is a second covered street. Lifts rising through this space recall the vertical movement of grain, which gave the complex its original purpose. Flooring details either retain existing turntables or hint at their historic location.

These spaces, together with lecture theatres, learning zones, canteen, exhibition areas and a roof terrace are robust, full of natural light and designed to be adaptable and raw. It is a stage for transformation, a framework of flexible spaces that can be orchestrated by staff and students where new interactions and interventions, chance and experimentation can create that slipsteam between disciplines, enhancing the student experience.

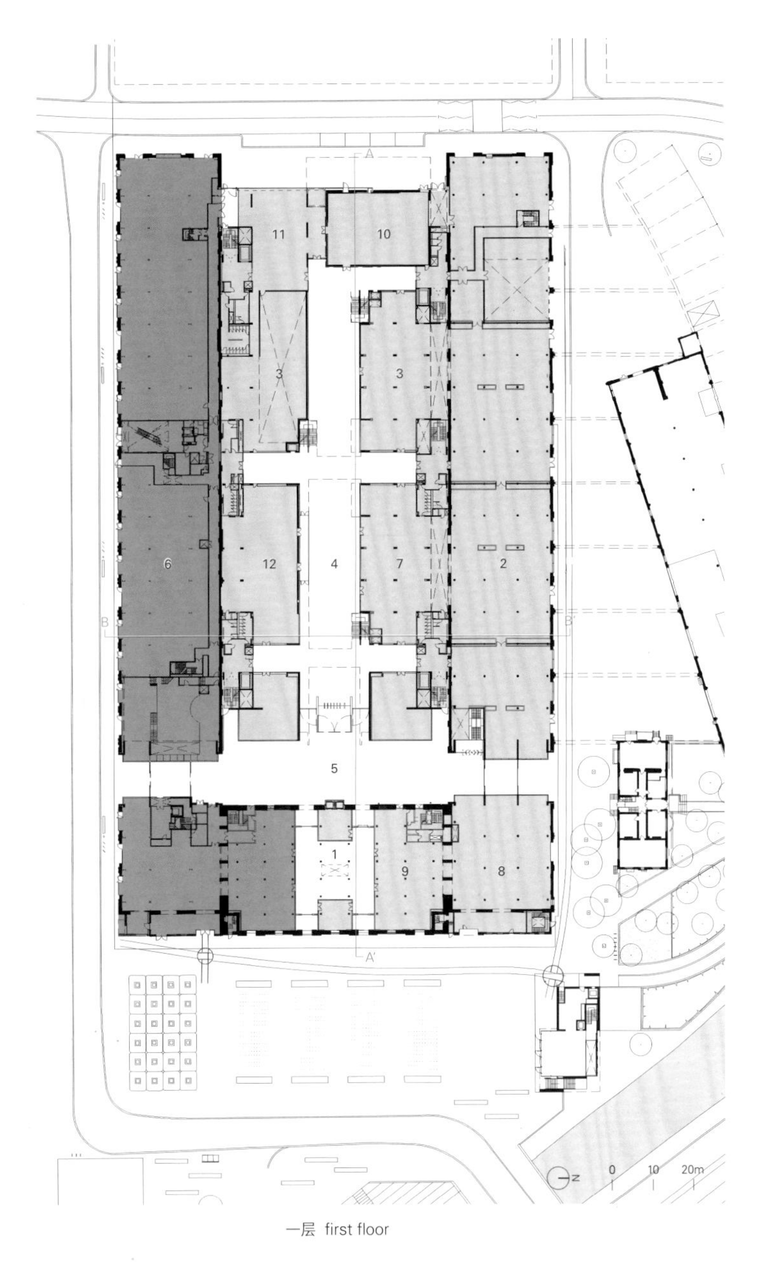

一层 first floor

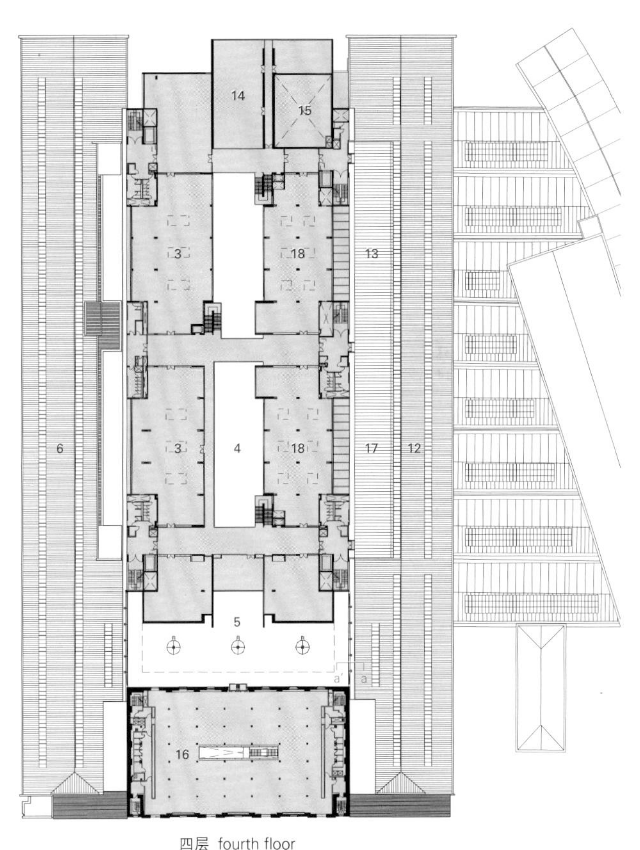

四层 fourth floor

1 谷仓建筑	10 剧院	1. granary building	10. theater
2 东部的货栈	11 剧院门厅	2. eastern transit shed	11. theater foyer
3 工作室建筑	12 阶梯教室	3. studio building	12. lecture theater
4 中央大街	13 露台	4. central street	13. terrace
5 东西侧连接区（公共通道）	14 舞蹈工作室	5. east-west link (public access)	14. dance studio
	15 塔台		15. fly tower
6 西部的货栈	16 图书馆	6. western transit shed	16. library
7 食堂	17 时尚工作室	7. refectory	17. fashion studio
8 Leathaby画廊	18 平面设计工作室	8. Leathaby Gallery	18. graphic design studio
9 接待处		9. reception	

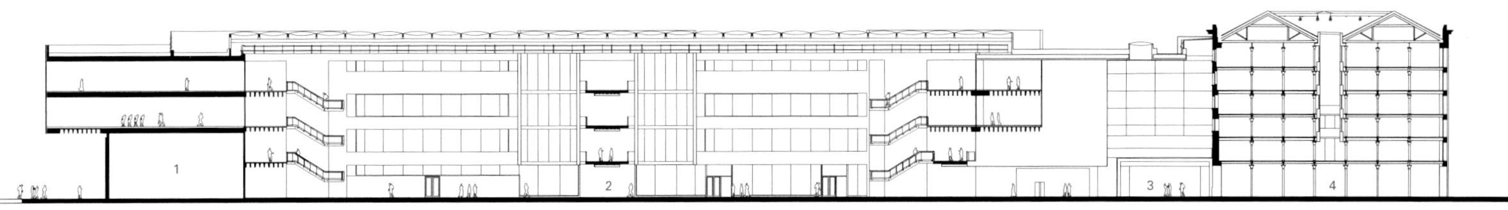

1 表演中心 2 中央大街 3 东西侧连接区(公共通道) 4 谷仓建筑
1. performance center 2. central street 3. east-west link(public access) 4. granary building
A-A' 剖面图 section A-A'

1 西部的货栈 2 工作室建筑 3 中央大街 4 东部的货栈(工作间/工作室) 5 露台
1. western transit shed 2. studio building 3. central street 4. eastern transit shed(workshop/studio) 5. terrace
B-B' 剖面图 section B-B'

项目名称：UAL Campus for Central Saint Martins College of Arts and Design
地点：Granary Building, 1 Granary Square, London, N1C 4AA, United Kingdom
建筑师：Stanton Williams / 承租人：University of the Arts London
结构工程师：Scott Wilson / 环境与M&E工程师：Atelier 10
照明设计师：Spiers and Major
工料测量师/雇主：Davis Langdon
景观建筑师：Townsend Landscape Architects
立面设计顾问：Arup Facades Engineering / CDM协调员：Scott Wilson
用地面积：27,700m² / 总建筑面积：14,400m² / 有效楼面面积：40,000m²
施工时间：2008.1—2011.4
摄影师：©Hufton+Crow(courtesy of the architect)(except as noted)

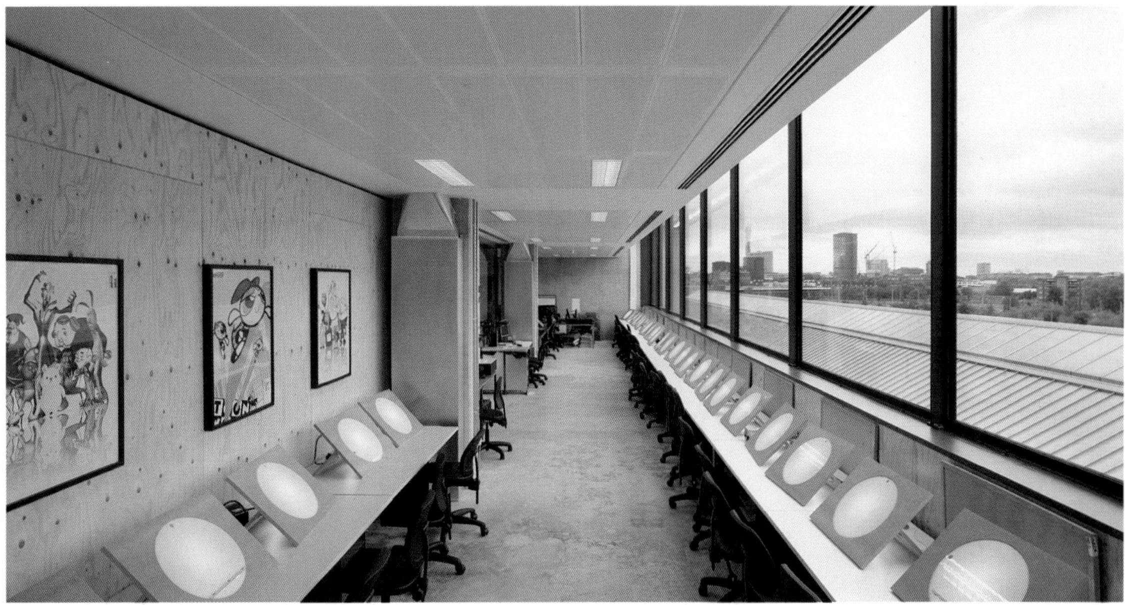

昆士兰大学现代工程学院大楼

Richard Kirk Architect + HASSELL

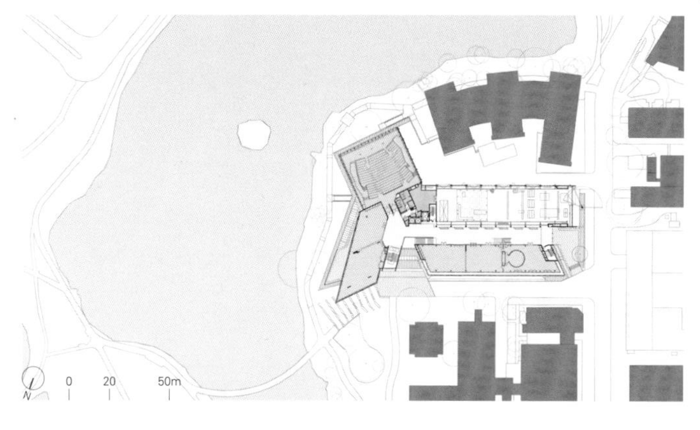

2009年，在经过一场设计竞赛后，Richard kirk建筑师事务所和HASSELL设计公司被任命为现代工程学院大楼（AEB）的建筑师，该项目耗资将达1.3亿美元。AEB项目旨在在昆士兰大学内建造一座建筑，在这处整合的工程景观内，师生们可以进行庆祝、合作、创造、学习、创新等活动，并使其持续发展。

通过全面的研究、研究生培训及本科生教育，AEB项目进一步推动了工程教育的转型。为了让游客和工作人员能够更清楚地看到流程和设备，建筑的结构要尽可能地开放。大面积的前厅用做成品展示，以激发人们关于制造过程和利用先进材料的兴趣。AEB项目涉及到高度复杂的建筑功能，其中包括土木工程学院和昆士兰先进材料加工制造中心所使用的研究设施、教学场所以及实验室和办公住宿场所。细心的规划能够克服任何潜在的问题，创造干净且富有创意的空间和流线，并且达到大学建筑所追求的高透明度和令人满意的互动及协作效果。

该建筑结构已被设想为一座"富有生命的建筑"——允许气候和结构方面的实时、有限性能监测，旨在使建筑成为一个终身学习的工具，以及课程学习的一部分。其布局和总体规划分为三个体量，均环绕着中央的垂直公共街道。平面之间的联系以及邻近的建筑为我们提供了一处灵活、活跃的学习空间，学习空间与实验室进行整合，以进行进步教学法。项目功能具有阐释性能，并且在水平和垂直方向上进行连接，以创造发现的意义和过程。AEB项目是一座绿星建筑认证的五星级建筑，通过简单的系统来实现，其中包括一座中庭（引入适度的空气和光线进入大楼）、具有高效性的立面、混合式通风系统、夜间冲洗系统和良好的采光层。

除了针对大型项目的独有能量战略之外，昆士兰本地木材的使用对于可持续发展来说也是十分重要的。建筑师尤其关注本项目中使用的当地木材在审美、环境及结构方面的优点的平衡，包括结构立面使用的玻璃及大跨度的胶合板屋顶桁架。这些项目策略由于与教育法有直接的联系，因此极具创新性，它们通过木材的可持续性能来展现结构，且保证了拥有500座席的礼堂的高质量的音质或音色。空间并没有使讲话产生回音——这促进了更密切的教授和学习体验。这种做法的目的是为了演示整座建筑内应用的木材在功能和美学上的潜力，从而带来可持续建筑的一个新标准，并且实现对当地可持续产业的承诺。

Advanced Engineering Building at the University of Queensland

In 2009, Richard Kirk Architect and HASSELL in Association were appointed as Architects for the $130 million Advanced Engineering Building(AEB) project, following a limited design competition. The AEB delivers on the University of Queensland's vision to create a building to celebrate, collaborate, create, learn, innovate and sustain, all within an integrated engineering landscape.

49
Advanced
Engineering
Building

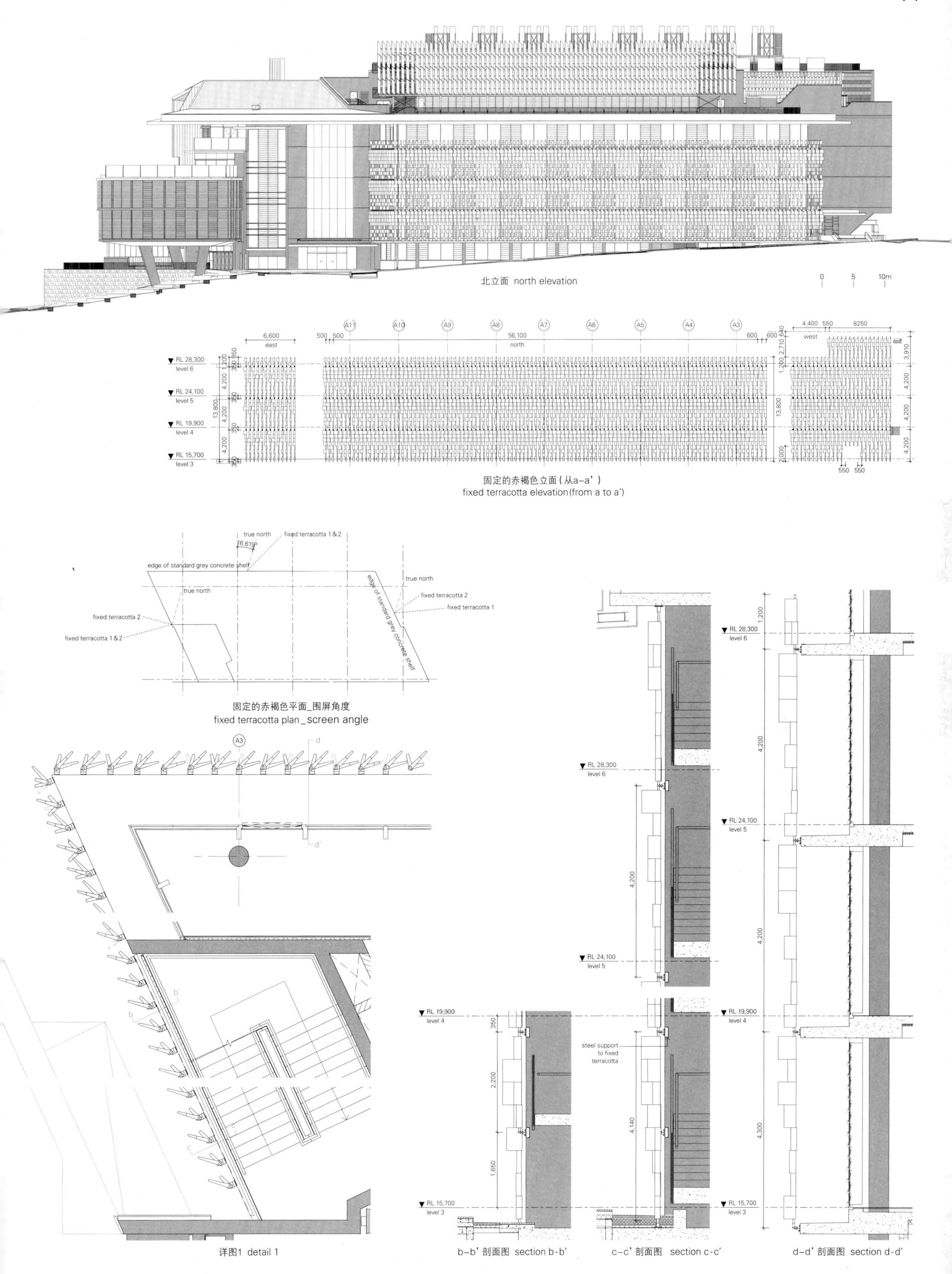

北立面 north elevation

固定的赤褐色立面（从a~a'）
fixed terracotta elevation(from a to a')

固定的赤褐色平面_围屏角度
fixed terracotta plan_screen angle

详图1 detail 1

b-b' 剖面图 section b-b'

c-c' 剖面图 section c-c'

d-d' 剖面图 section d-d'

The AEB facilitates the transformation of engineering education through all aspects of research, postgraduate training and undergraduate education. The structure of the building is as open as possible, to allow a high level of visibility of processes and equipment to visitors and staff alike, with a large foyer area to be used to showcase manufactured goods, stimulating interest in manufacturing processes and utilizing advanced materials. AEB involves a highly complex building program, which includes research facilities, teaching and learning spaces, laboratories and office accommodation for the School of Civil Engineering and the Queensland Center for Advanced Materials Processing and Manufacturing(AMPAM). Careful planning overcame any potential issues, generating clear and innovative spaces and circulation, and achieving the high level of transparency, interactivity and collaboration sought by the University.

The structure has been conceived as a "living building" – allowing real-time, finite monitoring of performance in climatic and structural terms. This is driven by the intention that the building becomes a lifelong learning tool and a part of the curriculum. The layout and general planning are integrated into three volumes that feed off a central and communal vertical street. Planning relationships and adjacencies provide and integrate flexible, active-learning spaces with laboratories allowing for progressive teaching. Program is located and displayed in an explanatory nature, linking the program horizontally and vertically creating a sense and process of discovery. AEB is a 5 Star Green Star As-Built certified building, achieved through the use of simple systems, including a central atrium to introduce tempered air and light into the building, highly efficient facades, mixed mode ventilations, night purging and excellent daylight levels.

The use of local Queensland timber became central to develop sustainability initiatives beyond the singular energy strategies typically targeted in large projects. Particular focus was placed on leveraging the aesthetic, environmental and structural strengths of using locally produced timber in the building – including structural facade glazing and long span glulam roof trusses. These strategies are also innovative in their direct link to pedagogy with the expressed structure utilizing timber for its sustainability credentials and contribution to acoustic quality or "color" in the 500-seat auditorium. The ambition for this space was also to allow for spoken word without amplification – an idea that promotes a more intimate teaching and learning experience. The purpose was to demonstrate the functional and aesthetic potential of timber across the building, resulting in a new benchmark for sustainable architecture and an ongoing commitment to local sustainable industry.

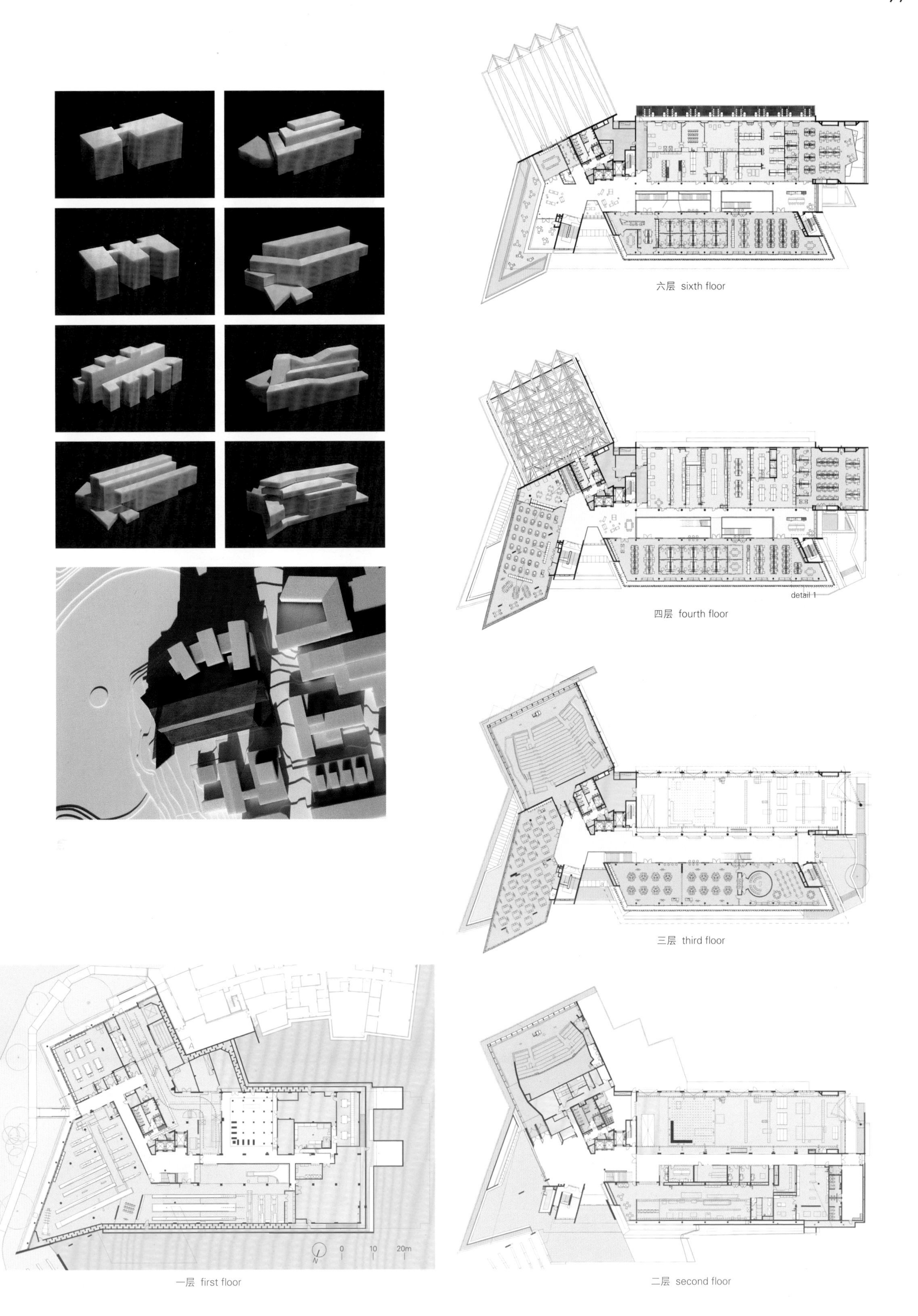

六层 sixth floor

四层 fourth floor

三层 third floor

二层 second floor

一层 first floor

项目名称：Advanced Engineering Building
地点：Brisbane QLD, Australia
建筑师：Richard Kirk Architect, HASSELL
项目团队：Richard Kirk Architect_Richard Kirk, Paul Chang, Yee Jien, Andrew Drummond, Fedor Medek, Jonathan Ward, Stephen Chandler, Adam Laming, Mitch Reed, Shane Willmett, Erik Sziraki, Matthew Mahoney, Lynn Wang, Grace Egstorf / HASSELL_Mark Loughnan, Mark Roehrs, Mark Craig, Joe Soares, Peter Hastings, Daniel Loo, Alison Hortz, Catherine Van Der Heide, Troy King, BP Loh, George Taran, Nguyen Luu, Mac Young, Fraser Shiers, Cheong Kuen, Amy Carrick, Greg Allis
结构/电气工程师：Aurecon
机械工程师：WSP Group
景观建筑师：HASSELL
承包商：WATPAC
总建筑面积：20,000m²
有效楼层面积：20,000m²
设计时间：2009—2010
施工时间：2011—2013
摄影师：©Peter Bennetts(courtesy of the architect)

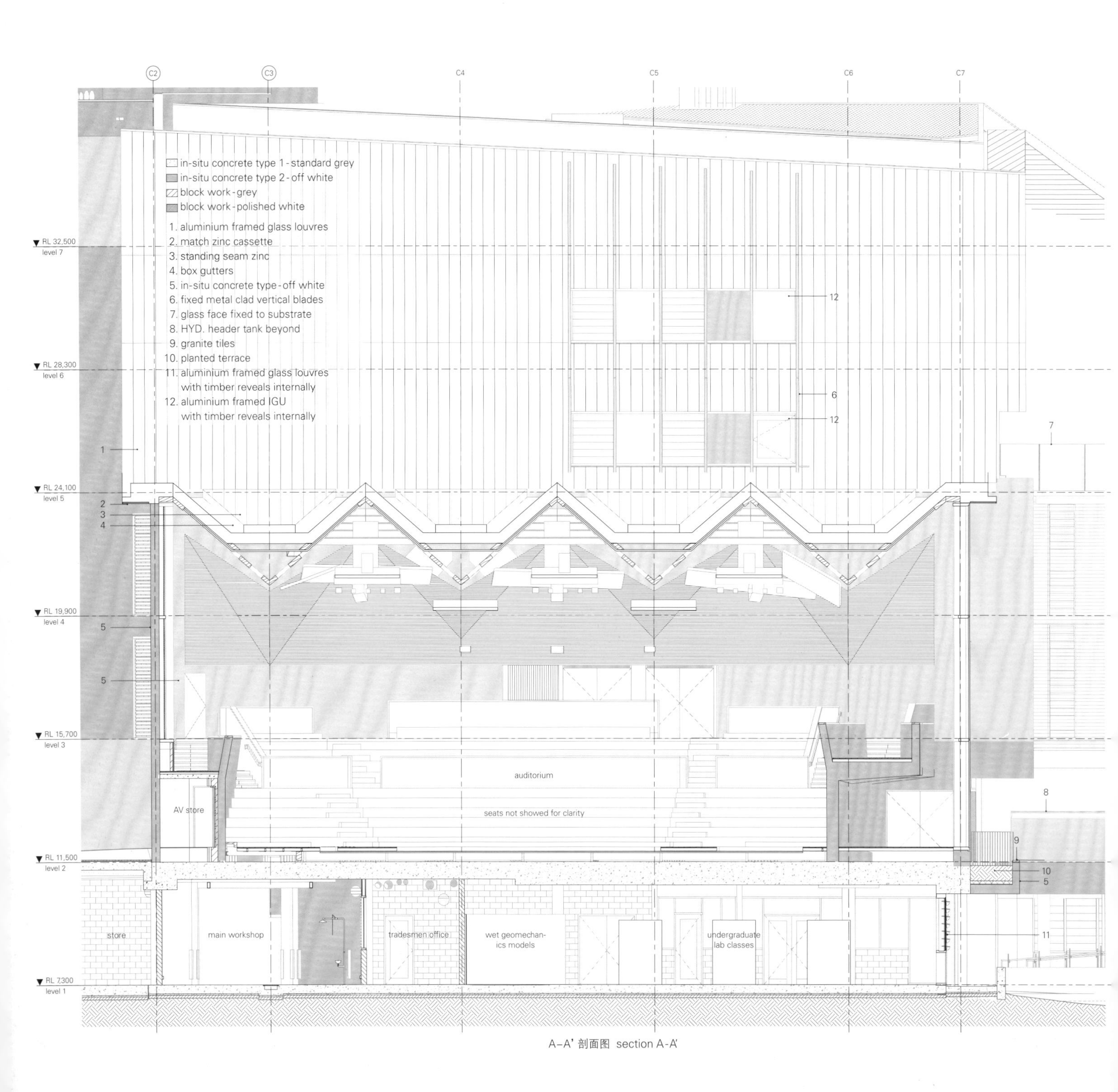

A–A' 剖面图 section A-A'

南丹麦大学柯灵校区

Henning Larsen Architects

低能耗建筑的领跑者

在丹麦，科灵校区是第一个能满足《2015建筑条例》中低能耗要求的大学建筑。其每平方米年耗能达36千瓦时。可持续发展战略融入到该建筑设计中：几何外形确保了每平方米都得到最大化的利用，而中庭的天窗确保了进入建筑的日光的均匀分配。这座建筑是一个为期三年的开发项目的一部分，该项目验证了提高混凝土的热性能，使制冷和制热系统所需的能耗得以降低的方法。为将混凝土的热性能进行优化，面板尽可能地暴露在阳光下，这种做法可以防止温度产生大波动，并且提高室内的建筑质量。此外，建筑师还实施了一系列不同的举措，以确保低能耗。

富有创意的立面

每时每刻，日光都在变化。因此，科灵校区配置了动态遮阳系统，能够进行调整以适应特定的气候条件和用户模式，并且沿立面提供最佳的光线以及舒适的室内环境空间。遮阳系统由大约1600个穿孔钢板形成的三角百叶组成。百叶嵌在立面表面，且能够调整以适应变化的光线及所需要的光照量。当百叶窗关闭时，它们会平铺于立面；而当它们从立面伸出，半开放或完全开放时，便会使建筑的外观变得富有表现力。遮阳系统安装了传感器，用来连续测量光照水平和热度，并且通过一个小电机来机械地调节百叶。

照明

南丹麦大学科灵校区以“需求为本”的照明为特色。高效节能的LED照明已应用于整座建筑中。

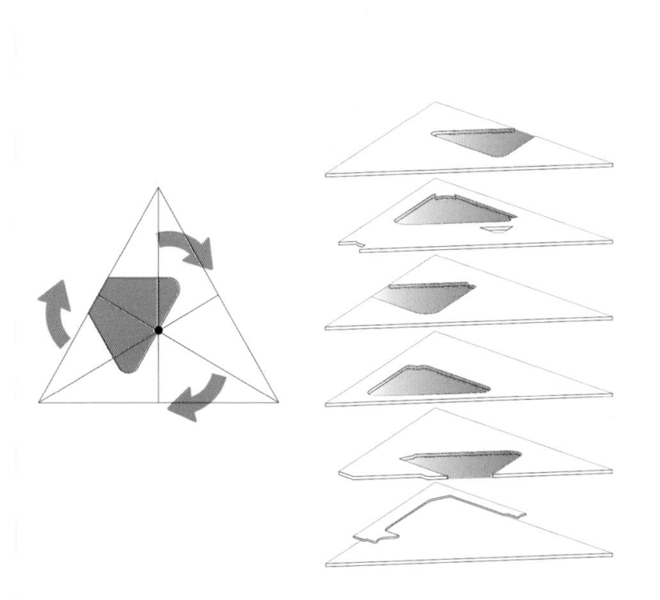

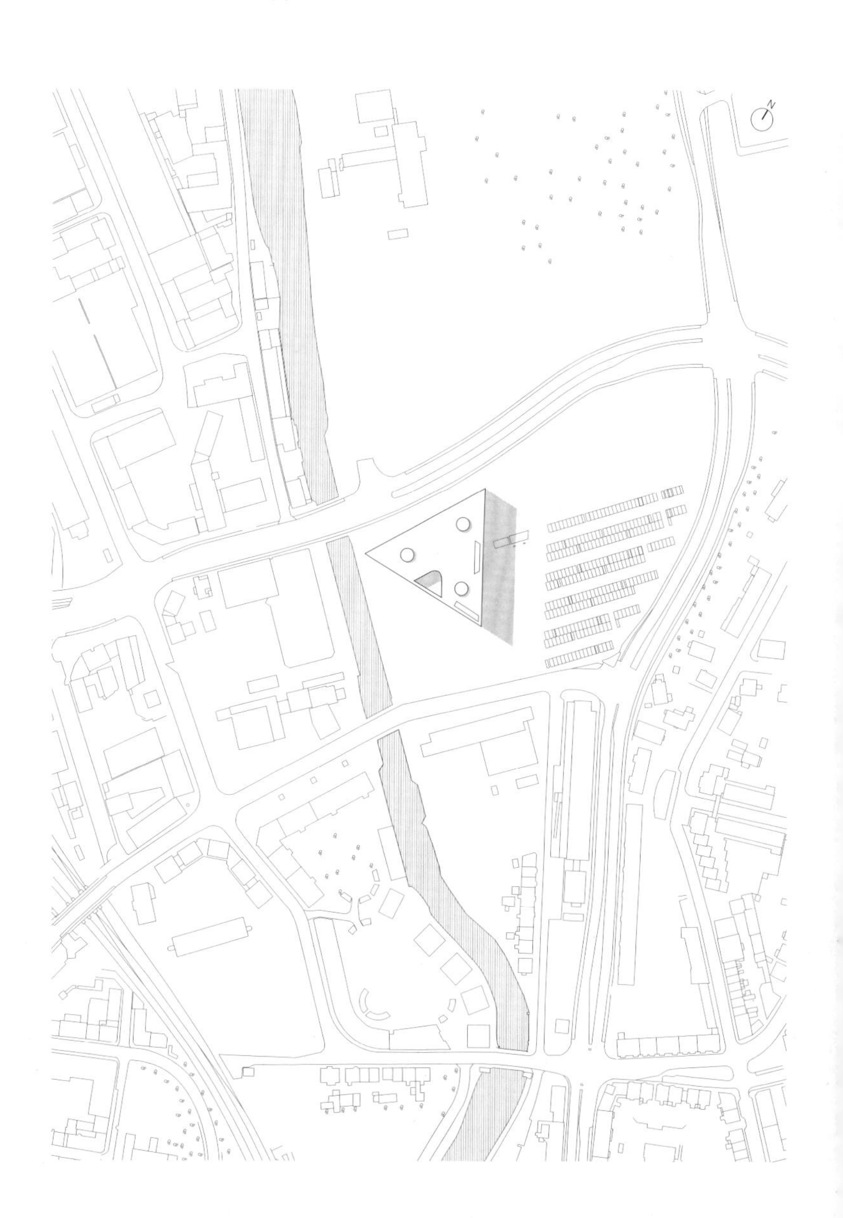

自然通风

低能耗的通风设计配有通风管和大型通风系统。新鲜的空气通过天花板上下流通，而开阔的中庭有助于空气流通。

作为制冷和制热系统的地下水井

两口地下水井有25m深。在深处，地下水的温度稳定，人们可以根据时间和需要，利用地下水对建筑进行加热和制冷。夏天制冷需求较高，温水储存在一个25m深的地下仓库中。当加热需求增高时，水流回转，仓库中的温水通过水泵送至上面。

太阳能电池系统

屋顶装有400m²的太阳能电池系统和20m²的太阳能加热板。

保温材料

建筑内还采用了标准方案，如安装在屋顶的、具有高保温性能和厚绝缘性能的三层玻璃。此外，真空保温材料已应用到窗体后方，这种材料的性能优于传统保温材料10倍，并且减小了窗体后方的厚度。

Kolding Campus of SDU

Front Runner in Low Energy Building

Kolding Campus is the first university building in Denmark to fulfill the strong demands for energy consumption described in the Building Regulations 2015. The energy consumption is 36 kWh/m²/year. The sustainable strategy has been integrated in the architectural design. The geometry ensures the best utilization of every square meter and the skylight in the atrium ensures a fair distribution of daylight to the entire building. The building has been part of a three-year development project which examines how thermal properties of concrete can be increased – and the energy consumption for heating and cooling thus reduced. In order to make optimal use of the thermal properties of concrete, the slabs are exposed where possible. This prevents large fluctuations in temperature and improves the indoor quality. Additionally, a series of different initiatives have been implemented in order to ensure the low energy consumption.

An Innovative Facade

The daylight changes and varies during the course of the day and year. Thus, Kolding Campus is fitted with dynamic solar shading, which adjusts to the specific climate conditions and user patterns and provides optimal daylight and comfortable indoor climate spaces along the facade. The solar shading system consists of approx. 1,600 triangular shutters of perforated steel. They are mounted on the facade in a way which allows them to adjust to the changing daylight and desired inflow of light. When the shut-

ters are closed, they lie flat along the facade, while they protrude from the facade when half-open or entirely open and provide the building with a very expressive appearance. The solar shading system is fitted with sensors which continuously measure light and heat levels and regulate the shutters mechanically by means of a small motor.

Lighting

SDU Kolding Campus features needs-based lighting. Energy-efficient LED lighting has been applied in the entire building.

Natural Ventilation

Low energy ventilation has been designed with air ducts and systems of large dimensions. Fresh air is blown in and down through the ceiling. The great atrium helps circulate the air.

Ground Water Wells as Cooling and Heating System

Two ground water wells, 25m deep, have been drilled. At the

depth, the ground water temperature is stable and the ground water could either warm up or cool down the building according to the time of year and the needs. In summer where the need for cooling is high, warm water is stored in a depot in 25m depth. When need for heating arises, the stream of water is turned around and warm water from the depot is pumped up.

Solar cell system

A 400m² solar cell system and 20m² solar heating panels have been installed on the roof.

Insulation

Standard solutions such as three-layered glasses with high insulation property and thick insulation on the roof have been applied. Furthermore vacuum insulation has been applied to the window backs. Vacuum insulation performs 10 times better than traditional insulation. This has reduced the thickness of the window backs.

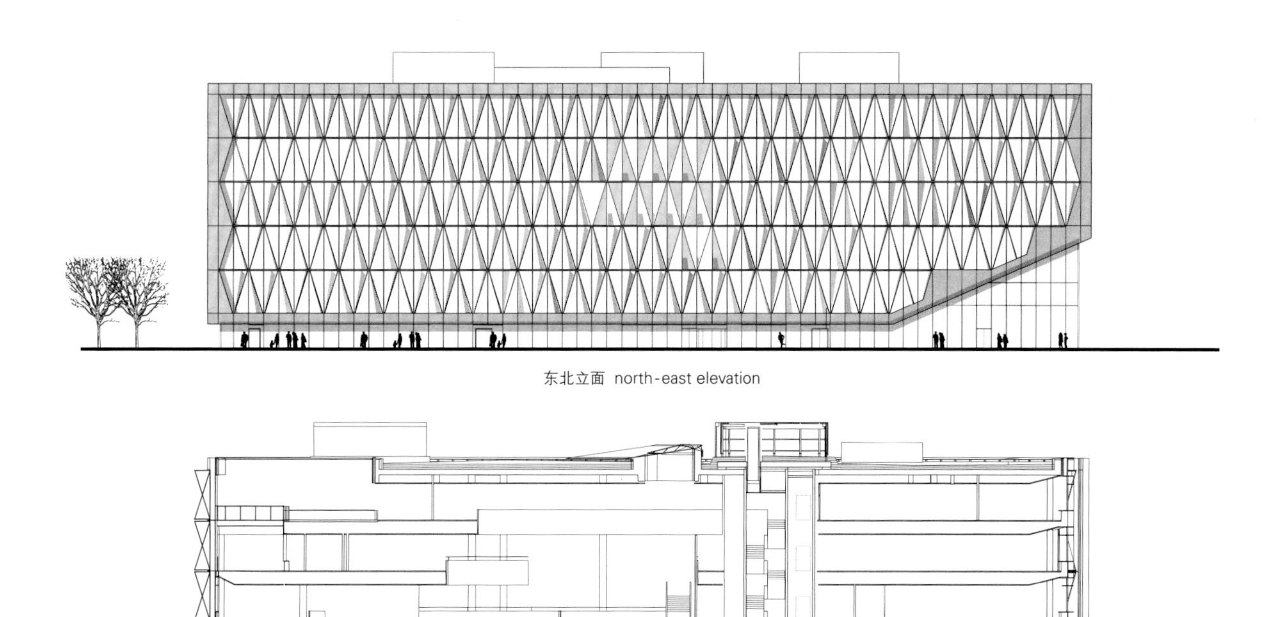

东北立面 north-east elevation

A-A' 剖面图 section A-A'

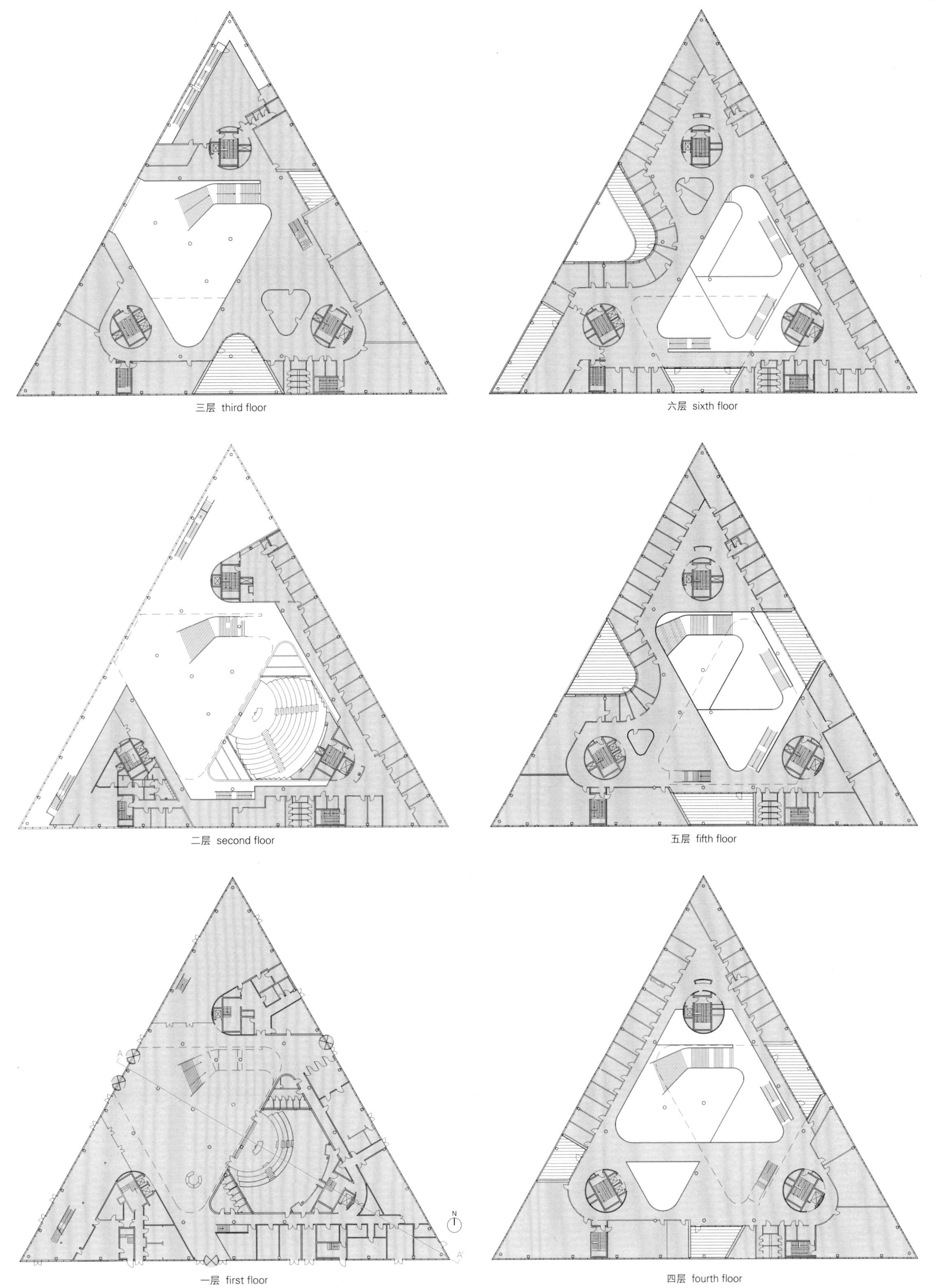

三层 third floor

六层 sixth floor

二层 second floor

五层 fifth floor

一层 first floor

四层 fourth floor

项目名称：SDU Campus Kolding / 地点：Universitetsparken 1, Kolding, Denmark / 建筑师：Henning Larsen Architects
景观建筑师：Arkitekt Kristine Jensens Tegnestue / 工程师：Orbicon / 甲方：The Danish Building & Property Agency
有效楼层面积：13,700m^2 / 竞赛时间：2008 / 竣工时间：2015
摄影师：©Jens Lindhe(courtesy of the architect) - p.106, p.107, p.108 top, p.111 top, p.112 / ©Martin Schubert(courtesy of the architect) - p.104~105, p.108 bottom, p.110, p.111 middle, bottom, p.113 / ©Jørgen Webert (courtesy of the architect) - p.103

增建与拆除

Addition and

在对老建筑进行改造并使它们适应新的用途时，需要保留什么，移走什么以及怎样添加新的元素便成为了关键问题。本章中所列的五座建筑全都对这些问题进行了处理，一些只是轻微地进行了改造，有一两处简单的移动，使整座建筑及其转运方式产生变化。其他建筑则是大胆地嵌入一些结构，以改变它们的影响。mlzd建筑事务所设计的位于瑞士拉波斯维尔-乔娜市的"两面神"建筑、伦佐·皮亚诺建筑工作室设计的哈佛艺术博物馆、Manuelle Gautrand建筑事务所设计的位于法国locon的Comedie de Bethune——国家戏剧院，Flores&Prats事务所和Duch-Pizá 事务所设计的位于马略卡岛帕尔马的卡萨尔·巴拉格尔文化中心以及noAarchitecten事务所设计的哈塞尔特大学都讲述了一些改造所带来的挑战，并且提供了一系列的处理方法。

In renovating old buildings and adapting them to new uses the question of what to keep, what to remove and how to go about adding new elements is key. These five buildings all tackle this issue, some very subtly, with one or two simple moves which transform the entire building and the way it works, and others intervene more boldly to effect their changes. mlzd's Janus building in Rapperswil-Jona in Switzerland, Renzo Piano Building Workshop's Harvard Art Museums, Manuelle Gautrand Architecture's Comedie de Bethune – National Drama Theater in Locon, France, Flores & Prats and Duch-Pizás Casal Balaguer Cultural Center in Palma de Mallorca and noAarchitecten's Hasselt University illustrate some of the challenges of renovations and offer a range of approaches to tackling them.

Subtraction

在什么值得保留以及为什么其对整个改造项目至关重要，从而使整座建筑适应新用途的问题上，总是有一个疑问的存在，即怎样设计这些增建结构。这里展示的建筑全部再现了老建筑的重要改造，但是建筑师们采取了不同的策略来处理，一些建筑利用一两个微小的改动来产生变化，其他则是大胆地对其进行处理。作为一组项目，它们讲述了这类建筑产生的关键问题，并且提供了不同的答案。

在大多数历史建筑的改造项目中，总是有一些小问题需要我们来回答，即需要保护什么，摒弃什么。"两面神"项目是mlzd建筑事务所的获奖竞赛项目，它连接了瑞士拉波斯维尔−乔娜市苏黎世湖边上的两座拥有700年历史的老建筑。这两座老建筑沿小镇城墙而建，起初的用途是堡垒，与住宅建筑相邻，后来在1943年与相连的建筑一起被改造成为博物馆。但在2002年，这些建筑便不仅仅作为博物馆使用。

在本案中，关于拆除哪座建筑的决定非常明确，20世纪40年代建造的、起连接作用的建筑（带有20世纪60年代建造的模仿中世纪风格的半面木质立面）必须移除，然后在原址上建造一个新的连接结构。虽然新结构的规模非常小，但是其用途是致力于解决现有的博物馆产生的多种问题。对于任何一座用途发生改变的建筑来说，老建筑的空间设计理念并不是设计成一座博物馆，也没有像可能的那样运作起来。雪上加霜的是，这座建筑曾经进行过一次翻新，再一次与老建筑的特征重合。而最终导致的结果是，到2002年，贯穿建筑的路线变得混乱且复杂，且存在数个不符合建筑条例和安全法规的建造实例。

这种新建的连接结构使建筑师在开始处理博物馆的空间问题方面，有了很大的自由度。但是，在该保留什么和保留原因的方面，该结构无法给予一些答案。要使一座建筑适应其用途，对历史肌理的改造可能是必要的，但是在商讨是否拆除某些结构的时候，历史肌理却强烈地暗示了施工品质可能与空间质量同等重要。就是否拆除来说，这一定是一

In the question of what is worth keeping and why is central to any renovation project, but in adapting them to new uses, there is also the question of how to go about making additions. The buildings shown here all represent significant transformations of old buildings, but the architects have taken different strategies to achieve them, some effecting the changes with one or two simple, subtle moves, others taking a bolder tack. As a group, they illustrate some of the key questions raised by these sorts of projects and provide a variety of answers.

In most renovations of historical buildings there are delicate questions to answer about what to preserve and what to do away with. The Janus building was mlzd's competition winning entry for a new connection between two 700 hundred year old buildings in Rapperswil-Jona in Switzerland, on the edge of Lake Zürich. The two started out as a fortified tower along the town's walls and adjacent residential building, but were converted into a museum, completed with connecting building, in 1943. By 2002 this collection of buildings was no longer up to the task of properly housing the museum.

In this case, the decision of what to demolish may have been straightforward – the 1940s connecting building, with its 1960s mock-medieval half-timbered facade had to go, with a new connection built in its place. Although the new building is relatively small, it was put hard to work to resolve the existing museum's many problems. As with any building where there is a change of use, the spaces of the old building weren't designed with a museum in mind and didn't work as well as they might. This was compounded by the fact that the building had already been renovated once, another common characteristic with very old buildings. The ultimate result was that by 2002 routes through the building had become confusing and convoluted and there were numerous instances where the building did not comply with building regulations and safety codes.

This creation of a new connector afforded the architect a lot of freedom to begin to tackle the spatial issues of the museum, but it does also suggest some answers to the question of what should

照片提供：Drawing©mlzd

照片提供：©Renzo Piano Building Workshop (Nic Lehoux)

哈佛艺术博物馆，马萨诸塞州，美国
Harvard Art Museums in Massachusetts, USA

个异常艰难的决定。例如，18世纪的精美雕刻嵌板，优于战后时期廉价的嵌入结构，也胜过略显尴尬的服务生空间。

mlzd设计的嵌入结构拥有其前身缺乏的品质和存在感。多面的立面覆有发光的穿孔金属板，能够让光线进入到室内，但是却无法传递视野到室外。新建筑的大部分室内空间用来设置可顶部采光的大型楼梯，这些楼梯以一种十分优雅的方式解决了早期存在的问题，即两座建筑的楼层没有位于同一平面上。

伦佐·皮亚诺建筑工作室设计的哈佛艺术博物馆是一座新建筑，它强烈地维护着自身的存在感。项目书要求在大学内建造三座博物馆，即Fogg博物馆、Bush–Reisinger博物馆和Arthor M.Sackler博物馆。它们都位于同一座建筑内，即一座原有的、20世纪20年代乔治亚时代的修复建筑内。此外，博物馆还扩建了其设施。其内容纳艺术学习中心、检测实验室、300个席位的礼堂、用于举办公共活动的扩建设施，以及三座博物馆的住宿区，这意味着原有的建筑需要大面积地扩建。看起来非常坚固的大型体块面向普雷斯科特街，且材质为玻璃和钢的屋顶结构非常状观，位于新建筑的上方，它们共同为这些要求提供了空间。

新建筑与原有建筑相联系，其庞大的规模改变了二者的关系，也许还减少了依附于老建筑的可能。从某种程度上来说，新建筑能够独立地设置功能，并且其内的大部分空间不会与老建筑有直接的连接，所以很多室内空间看起来相当的不引人注目。也许只有建筑顶层右侧区域内的保护区（位于玻璃屋顶之下，俯视着20世纪20年代建造的建筑的庭院）能够将新建筑与老建筑连接起来。

Manuelle Gautrand建筑事务所对法国Locon的国家戏剧院（Comedie de Bethune）进行了一系列的扩建，这些新建建筑使原有建筑在更大程

be kept and why. Although alterations to historic fabric may be necessary to make a building fit for purpose, it suggests strongly that quality of construction may be just as important as quality of resulting space when making the call as to whether to demolish something. It must be a harder decision to remove, for example, 18th century beautifully carved paneling, than a cheaply built post-war addition and its awkward attendant spaces.

mlzd's intervention has a quality and a presence that its predecessor lacked. The faceted facade is clad in gleaming perforated metal, letting light penetrate to the interior, but giving away little to the outside. Much of the interior of the new building is given over to the large top-lit staircases which elegantly resolve the earlier issue that the floors of the two buildings were not at the same levels. With the new connection taking on this subservient function, the other two buildings can be dedicated more fully to use as exhibition spaces.

Renzo Piano Building Workshop's Harvard Art Museums new building however, asserts itself strongly. The brief called for the university's three museums, the Fogg, Busch-Reisinger and the Arthur M. Sackler, to be united within one building, an existing 1920s Georgian revival building, and to extend their facilities. The art study center, conservation lab, 300 seat auditorium and enlarged facilities for public events, together with the accommodation of the three museums meant that the existing building needed to be enlarged significantly. The large solid-looking block facing onto Prescott Street and the imposing glass and steel roof structure which sits atop the new buildings, offer space for some of these.

The sheer size of the new building relative to the existing one changes their relationship and perhaps makes it less likely that it would defer to the old one. To a certain extent the new building can do its own thing and certainly a large proportion of the spaces in the new part do not come into direct contact with the old building, so that many of the interior spaces seem fairly oblivious to it. Perhaps only the conservation areas right at the top of the building, under the glass roof and peering down into the court-

照片提供：©Manuelle Gautrand Architecture (Luc Boegly)

Comedie de Bethune——国家戏剧院，Licon，法国
Comedie de Bethune – National Drama Theater in Locon, France

度上变得矮小，并且设法在不同阶段建立起一种有趣的关系。在这里，虽然老建筑占有绝对的主导地位，但是它却要"小心翼翼"地融入到新扩建的建筑群中。老建筑的理念分为不同的部分，但是却是一个统一的整体。这在其平面和剖面中均有所体现。后台、礼堂、前厅以及彩排空间均各具特色，且位于三座不同的建筑之内，但是它们却和谐地坐落在一起，人们在看图纸的时候，根本无从得知这一事实。

这三座建筑中最古老的一座为20世纪30年代建造的影院，当它面临被改造成一座剧院时，只有立面保存下来。1994年举办的设计竞赛保留了礼堂和后台建筑的结构，而新部分呈弯曲状，覆在原有的立面之上，以模仿其原有的形式，并且漆成黑紫色，以和影院的灰白石材形成实时对比。场地一角的建筑进行了拆除，最后有可能增建一个新的入口和彩排空间。

建筑师对历史建筑进行了处理，使其以一个有趣且非比寻常的面貌展现出来。而最新的建筑则受到了不同的待遇，直线形外形，覆有黑色金属板，并且复制了紫色建筑的菱形图案。然而，最引人注目的还是影院的立面，建筑师决定让老建筑随着时间的推移，其颜色将逐渐淡化，最后融入整个大整体中。因此，20世纪90年代建造的建筑的紫色和黑色图案延伸至整个立面，并将其覆盖。纵观整个建筑群，新扩建的机构与老建筑一起，似乎完成了建造令人舒适的住宿区的目标。

Flores&Prats事务所和Duch-Pizá事务所联合设计的卡萨尔·巴拉格尔文化中心位于西班牙巴利阿里群岛的马略卡岛上的帕尔马，在这里，大量的新老建筑紧紧地交织在一起。该设计非常与众不同，它从老结构中借鉴细节，同时使其富有现代气息，来为建造大型空间、屋顶通道，以及一座动态的中央楼梯提供机会。

yard of the 1920s building have this connection between the old and new parts.

Manuelle Gautrand Architecture's series of extensions to the Comedie de Bethune, the National Drama Theater in Locon, France, dwarf the original building to an even greater degree, but have managed to establish a playful relationship between the different phases. Here it is definitely the older building that calls the tune, and yet it has been carefully assimilated into the newly enlarged complex. The sense of this building being made of different parts but creating a very unified whole can also be seen in the plans and sections. The backstage, auditorium and front of house and rehearsal space have different characters and are in fact in three different buildings, but they sit together so comfortably that you wouldn't necessarily know this from looking at the drawings.

The oldest building of the trio is a cinema from the 1930s, although when it came to convert it into a theater, only the facade was kept. A competition in 1994 saw the construction of the auditorium and backstage buildings, the newer part curving up and over the older facade, mimicking its form and painted dark purple, which at the time contrasted with the pale stone of the cinema. With the demolition of the building at the corner of the plot it was finally possible to add a new entrance and rehearsal space.

The architects have an interesting and unusual take on working with this historical building. The latest building has been treated quite differently, with its rectalinear shape and black metal panels, though they copy the rhombus shape of the purple building's pattern. Most notable is the cinema facade however; the architects decided that with time, the old building should be gradually erased, subsumed into the larger whole, and the purple and black pattern of the 1990s building has been extended to cover its facade. Looking now at the entire complex, the new additions seem to have reached a very comfortable accommodation with their older counterpart.

In architects Flores&Prats and Duch-Pizá's Casal Balaguer Cultural Center in Palma de Mallorca in the Spanish Balearic Islands, there is a much greater interplay of new and old. The design takes a very

Drawing©Manuelle Gautrand Architecture

照片提供：©Flores & Prats + Duch-Pizá (Adrià Goula)

卡萨尔·巴拉格尔文化中心，帕尔马，马略卡岛，西班牙
Casal Balaguer Cultural Center in Palma de Mallorca, Spain

原有的建筑是一座14世纪建造的住宅，在200年后的16世纪进行了改造和增建，之后又于18世纪做了同样的处理，现在，这座建筑被改造成为一座博物馆和文化中心。经过几个世纪的发展，这座建筑成为一座复杂的建筑，或者至少可以说，建筑师能够有效地开发这座建筑。原始建筑和扩建结构的不同几何外形之间的交汇处新建了主入口，而中心处，流线与照明相结合，使建筑内部指引更加明确，也更加直观。空间的复杂外形是其周围的原有建筑直接作用的结果。

建筑的其他区域内也存在着类似的微妙举动，即将空间巧妙地改造成更适合使用的文化中心的一部分。新建的木屋顶结构与老屋顶结构类似，是马洛卡区典型的宫殿屋顶，不同之处便是新梁的厚度，这种厚度使其间的跨度是老梁之间的跨度的两倍，能够为二层的图书馆创造更加宽敞的空间。建筑斜屋顶之间的露台提供了开放的空间，游客可在其间阅览图书。这个新设计与原来的建筑紧密结合，且通过几个细致的改动，将住宅改建为公共文化中心。

本系列的最后一个案例展示了一个相当极端的、对一座老建筑进行改造并使其适用于新用途时所面临的挑战和机遇。NoaArchitecten事务所设计的哈塞尔特大学位于比利时的东部，过去曾是一座监狱。这两种建筑类型都需要非常特殊的、与众不用的空间，与许多文化或者家用空间（能够经常融入不同类型的建筑内）不同。对监狱进行实体改造是建筑师所要面临的第一个挑战。第二挑战则是这座建筑的象征意义。该学校也渴望能够建造一座位于市中心的、与众不同的建筑。从某几个方面来说，这座于19世纪50年代建造的监狱是符合要求的。但是从其他意义来说，它又存在着问题；正如建筑师所言，"一座开放的大学真地能进入一座象征权力工具的大楼吗？"

different attitude. It borrows details from the old construction, but modernizes them and creates new opportunities for larger spaces, roof access, and a dynamic central staircase.

The original building is a 14th century house, remodeled and added 200 years later in 1500, again in 1700 and now finally converted to a museum and cultural center. Centuries of growth had produced a building that was complex, to say the least, the architects were able to exploit this productively. The point at which the different geometries of the original building and its extensions meet became the site of the building's new main staircase. Incorporating circulation and light at this central point made navigating the building easier and more intuitive. The complex shapes of this space are a direct result of the existing buildings which surround it.

There are similarly nuanced moves in other parts of the building, deftly transforming the spaces into ones more appropriate for use as part of a cultural center. The new wooden roof structure is similar to the old one, typical of palace roofs in Mallorca. The difference is the thickness of the beams, allowing them to double the span of the older beams and creating a more generous space for the second floor's library. The creation of terraces amongst the buildings' pitched roofs offers open air spaces for visitors to read. The new design works very closely with the existing building and with a few carefully chosen moves is able to make the change from house to public cultural center.

The last building in this collection presents a fairly extreme example of the challenges and opportunities of renovating an old building and adapting it to a new use. NoaArchitecten's Hasselt University in the east of Belgium used to be a prison. Both of these building types require very specific, and quite different, kinds of spaces, unlike many cultural or even domestic uses for example, which can often fit into a wide variety of building. The physical conversion of the prison presented the first challenge. The second challenge was the symbolism of the building. The university was seeking a building in the center of the city with a distinctive presence and in several respects the 1850s prison fitted the bill. In

照片提供：©noAarchitecten

哈塞尔特大学的学习中心模型
study models of Hasselt University

全新的法学院由三座不同的建筑构成，包括两座容纳办公室和教室的新建筑，它们的风格和形式都与老建筑形成强烈的对比。此外，还有这座监狱。乍一看人们很难分辨哪个是新建的结构，从空中的角度来看，星状的监狱外形非常鲜明。庭院上方的带有绿色屋顶的建筑完全地改变了牢房以及监狱翼楼的属性。宽敞的走廊延伸至原始建筑的室外和庭院（设置了大型空间，如监狱内无法设置的礼堂），而原来的牢房成为办公室和会议室。甚至中央的圆形监狱空间也被改造成为一处重要的定位和聚集点。

这座建筑巧妙地整合了改建所带来的机遇和限制。虽然本文所展示的其他建筑以一种更温和的形式对其进行了处理，但是这些问题仍然具有普遍性。物理变化和用途变化改变了一座建筑与周围建筑的关系。在哈塞尔特大学，监狱的墙体是为了囚困犯人，但是现在它呈现了一个全新的意义，即大学独有的、将全世界的所有其他空间排除在外。建筑师没有建造一座抬高的屋顶花园，以降低围墙的高度，但是至少从里面看，这是一座低矮的围墙。

从可持续性的角度来看，对原有建筑进行重新利用是十分有益的。这意味着至少在西方的欧洲和北美地区，建筑师的工作负荷会增加一部分。相对于思考这些建筑引发的一些问题来说，更重要的是，我们应该思考保留什么，移除什么，原因，以及相对与于那些类似的现代建筑来说，历史建筑占据着什么样的地位。

other senses it was problematic; as the architects put it, "can an open university move into a building conceived as an instrument of power?"

The new law faculty is composed of three separate buildings – two new ones containing offices and classrooms, which in both style and form contrast strongly with the old prison, and then the prison itself. At first glance it can be difficult to discern the new additions, the form of the star-shaped prison being so graphically strong in the aerial view. The green-roofed buildings in the courtyard have completely altered the nature of the cells and wings of the prison. With spacious corridors pushed to the outside of the original building and the courtyard housing large spaces, such as the auditorium that the prison couldn't, former cells can become office and meeting rooms. Even the central panopticon space can be diverted to become a central orientation and gathering point.

This building neatly sums up the opportunities and constraints of such transformations. Though the other buildings presented here may have dealt with them in a more moderate form, these issues are common to all of them. The physical change and change of use alter a buildings' relationship to its surroundings. In Hasselt, the prison wall, designed to keep inmates in, takes on another meaning – the exclusivity of a university keeping the rest of the world out. The architects do not create a raised roof garden that reduces this barrier, at least from within, it's a low perimeter wall.

Reusing existing buildings is also beneficial from the point of view of sustainability, meaning that in Western Europe and North America at least, it is likely to become an increasing part of an architect's workload. Ever more important than to consider some of the questions raised by these buildings we should consider, what should be kept and what demolished and why, what is the place of a historic building with regard to its modern counterparts?

Alison Killing

Drawing©noAarchitecten

“两面神”项目

mlzd

拉波斯维尔–乔娜市博物馆的扩建和修复项目

命名为"两面神"的项目赢得了2007年举办的设计竞赛，并且赋予了拉波斯维尔–乔娜市博物馆和其公共性同等重要的新形象。该建筑的设计目的是为了吸引那些并没有因为对文化感兴趣而望而却步（市政区界限的原因）的公众的注意力，并且让博物馆和城镇成为富有吸引力的旅行目的地。拥有新建筑的该项目已经敏锐地融入历史城镇当中。北部的视野对城镇的整体视觉印象非常重要，所以未曾改变。建筑小心翼翼地融入到狭窄的城市中心街道所呈现的历史图片背景中。因为处在全新的地形中，并且设有高品位的青铜立面，所以这座建筑使当前的环境变得十分显眼，并且很容易被解读为现代博物馆建筑群的主要入口。

除了主要的入口，博物馆的两个部分，即布莱尼室和布莱尼塔因为新建筑也可以让残疾人进入的原因，从而受益。作为整个建筑群的新部分，"两面神"项目满足了所有现代的、全年服务式博物馆的建筑服务和操作需求，并且使遗产建筑证明其存在过以真实见证那个时代成为可能。出于同样尊重历史的态度，新建筑的外形也在原有建筑的侧立面的基础上开发出来。它的立面和屋顶设计使旧建筑保留的窗户和门都没有被切断。

新建的房间使博物馆的空间、经营和可能策划的范围都有所延伸。例如，一层就设有若干个功能。访客在进入建筑综合体内后，能够立即进入两层高的主室。考虑到它的中心位置和所展示的整个城镇的模型，这个主室是博物馆或城市旅行最合适的起点。如果要举办大型活动，这里还可以和前院或画廊层连接起来。

除了主室外，博物馆还提供了很多不同的房间，以适用不同的展览功能。例如，三层有一些窗户嵌入到墙体的房间，它们提供了非凡的视野。此外，四层沐浴着自然光，一方面，使人们能够看到布莱尼室的令人印象深刻的屋顶，这个屋顶在以前是看不到的；另一方面，它还提供了对当代元素进行现代化展示的背景。在一天以及季节更替当中，自然光照射的不同方式给建筑增加了更加有趣的维度。

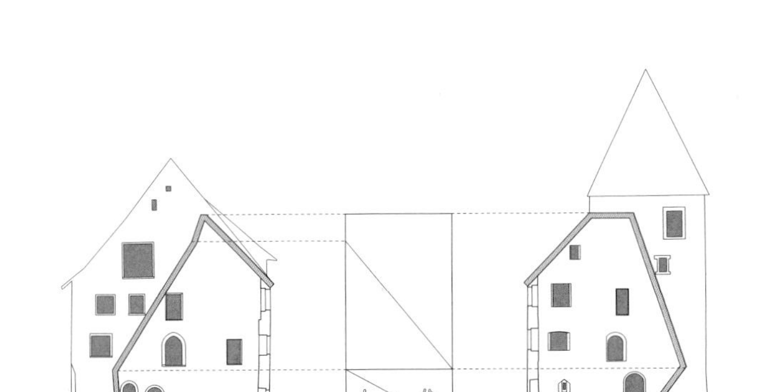

剖面示意图 section diagram

穿过屋顶的照明以及楼层间的光透射形成了内部空间和遗产建筑之间的鲜明对比。首先，这让人们在这个建筑综合体内很容易找到自己的路；其次，新建筑与老建筑相互制衡。因此，进入这座遗产建筑成为回到过去、历经大事件的时光之旅。新建筑对这一空间特性进行了参考，从而激发了游客的好奇心，刺激他们开始一次发现之旅。不同的观看视线让城市和博物馆展现出不同的方面和景色。它们令人们渴望在博物馆内走动，并且探知建筑，了解陈列的展览品。

Janus

Rapperswil-Jona Municipal Museum Extension and Renewal

The "Janus" project, which won a competition held in 2007, is giving the Rapperswil-Jona municipal museum a new profile commensurate with its public significance. It is designed to attract the attention of members of the public interested in culture without stopping at the municipal boundaries and presents the museum and the town as an appealing destination for excursions. The project to put up the new building has been sensitively integrated in the historic town. The view from the north, which is important for the overall visual impression of the town, is to remain unchanged. The building fits discreetly into the background of the historic picture presented by the narrow town-center streets. With the new terrain situation and the tasteful bronze facade, the building imposes a new emphasis on its immediate surroundings and can easily be read as the main entrance to a modern museum complex.

In addition to a new main entrance, the Breny House and Breny Tower parts of the museum now benefit from disabled access thanks to the new building. As a new part of the whole complex, "Janus" satisfies all the building-services and operational requirements of a modern, round-the-year museum operation and thus makes it possible for the legacy buildings too to justify their existence as authentic witnesses of their day and age. It is with this same respectful attitude that the shape of the new building has been developed out of the lateral facades of the old buildings. Its facade and roof have been designed in such a way that the existing windows and doors of the old buildings are not intersected anywhere.

东南立面 south-east elevation

西北立面 north-west elevation

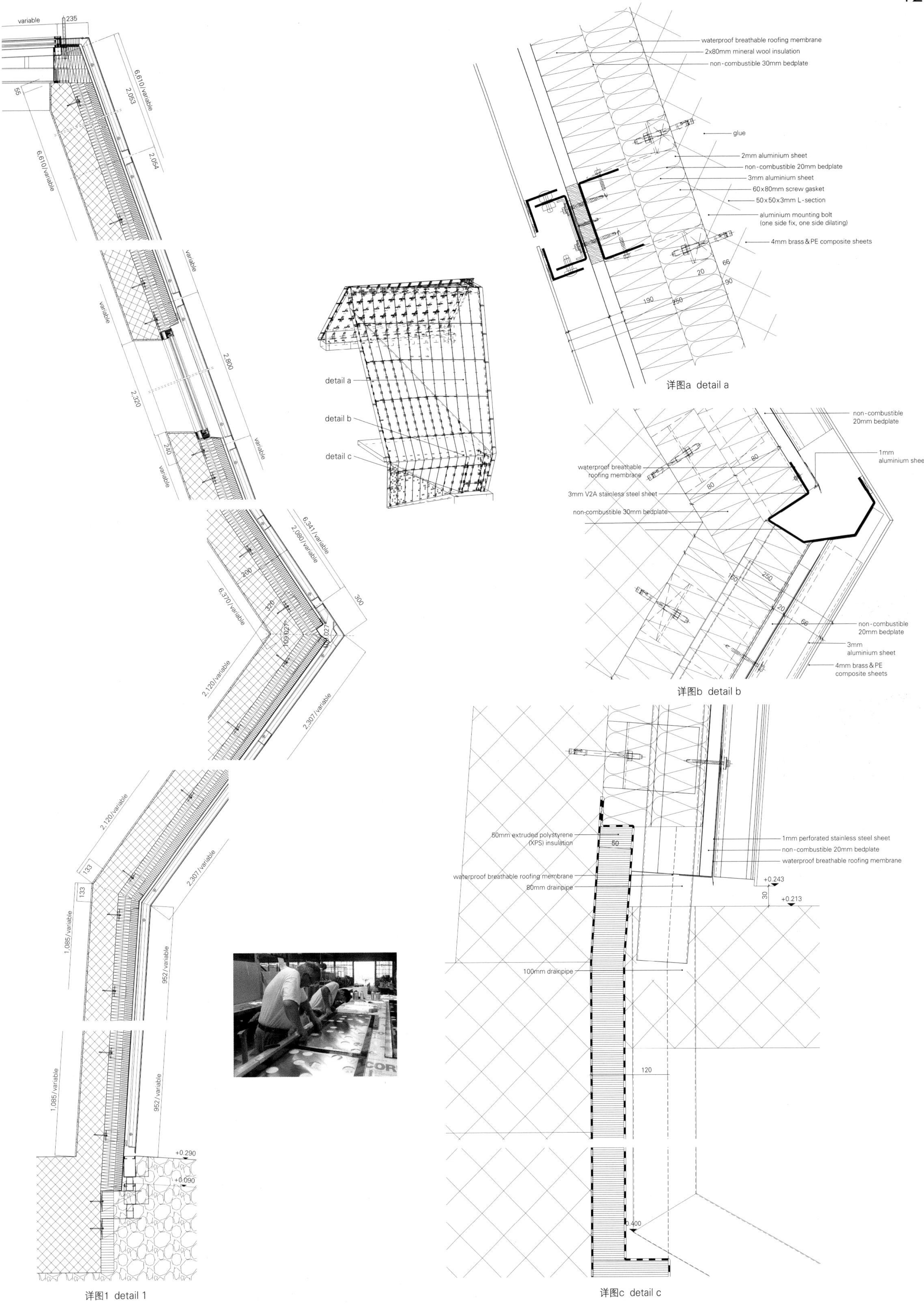

详图a detail a

详图b detail b

详图1 detail 1

详图c detail c

项目名称：Janus _ Extension and Renewal of the Rapperswil-Jona Municipal Museum
地点：Rapperswil-Jona, Switzerland
建筑师：mlzd
项目团队：Pat Tanner, Andreas Frank, Monica Udrea, Daniele Di Giacinto, Carol Hutmacher, Claude Marbach, Roman Lehmann, Regina Tadorian, Réne Robeck, Julia Wurst, Beat Junker, Frederike Kluth
甲方：Ortsgemeinde Rapperswil-Jona Stadt Rapperswil-Jona
表面面积：old buildings _ 290m² / new building _ 80m²
有效楼层面积：old buildings _ 820m² / new building _ 170m²
新建筑立面面积：200m²
造价：CHF 5.8 million
竣工时间：2011
摄影师：©Dominique Marc Wehrli

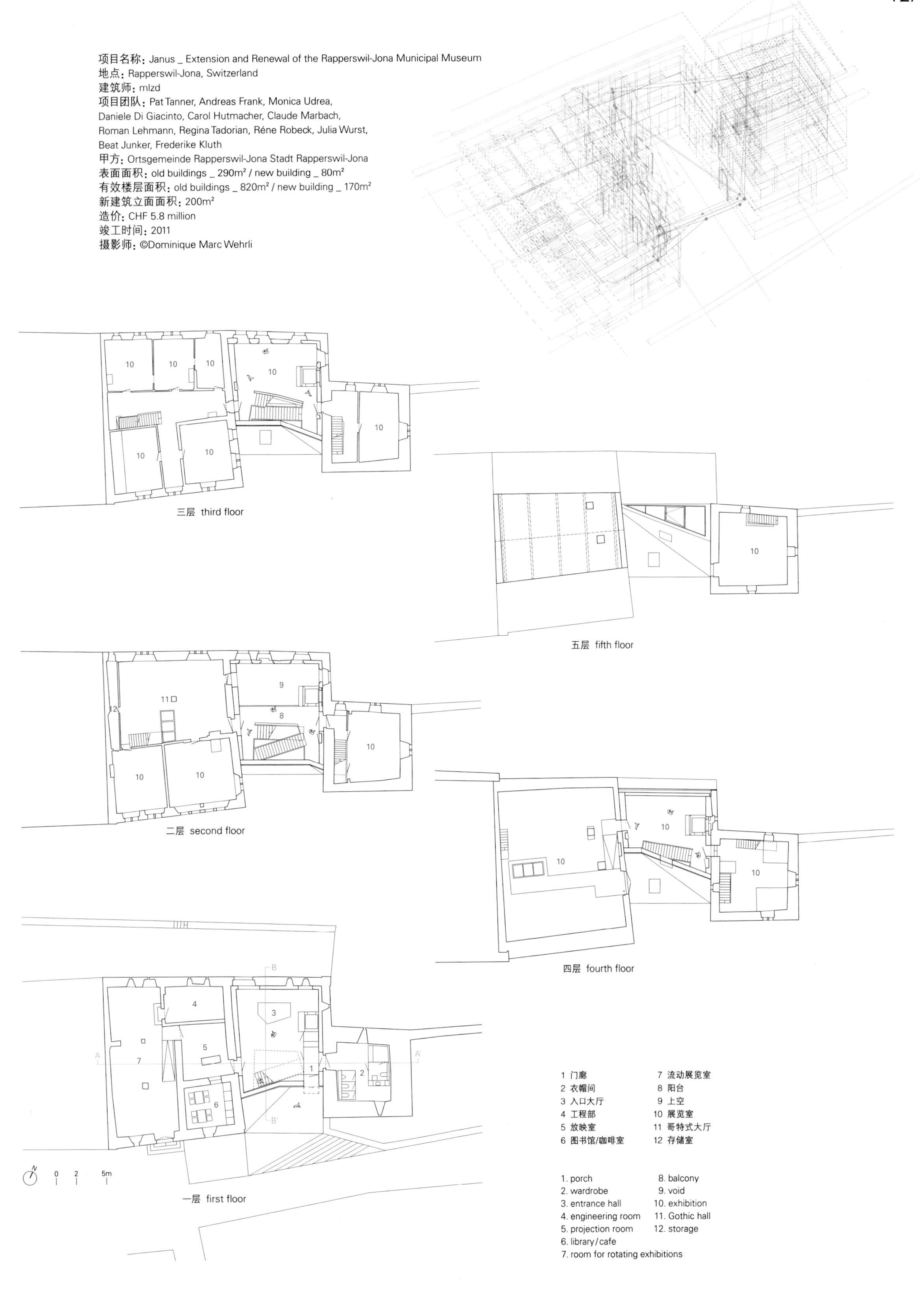

1 门廊	7 流动展览室
2 衣帽间	8 阳台
3 入口大厅	9 上空
4 工程部	10 展览室
5 放映室	11 哥特式大厅
6 图书馆/咖啡室	12 存储室

1. porch	8. balcony
2. wardrobe	9. void
3. entrance hall	10. exhibition
4. engineering room	11. Gothic hall
5. projection room	12. storage
6. library/cafe	
7. room for rotating exhibitions	

F

The newly created rooms are extending the museum's spectrum in terms of spaces, operations and the possibilities available to the curator. The ground floor, for example, fulfills several functions. After entering the complex, visitors immediately move into the main room, which is two floors high. Given its central position and the model of the town on display there, this room is a suitable starting point for conducted tours of the museum or the town. For prestigious events, it can be combined with the forecourt or the gallery floor.

Many different rooms are also available in addition to the main one and are appropriate for a variety of exhibition purposes. One example is the third floor, where the rooms have had plenty of windows incorporated in their walls, affording marvelous views. Another is the fourth floor, which is drenched in bright light, making it possible, on the one hand, to admire the impressive roof timbering of Breny House, which had previously not been visible, and, on the other hand, to provide a setting for a modern presentation of contemporary contents. The way that different types of natural light are brought into play adds another interesting dimension to the building in the course of the day and the succession of the seasons.

Illumination of the building through its roof and the transmission of light from floor to floor deliberately create a stark internal contrast with the legacy buildings. Firstly, that makes it easier for people to find their way around the whole complex and, secondly, the new is clearly offset against the old. Stepping into the legacy buildings thus becomes an eventful journey in time, back into the past. Thanks to spatial references of this nature, the new building kindles visitors' curiosity and stimulates them to set out on this journey of discovery. Various direct lines of vision show up the town and museum in unexpected perspectives and vistas. They create the desire to move around in the museum and to get to know the buildings and the exhibitions on display in them.

1 展览室 2 存储室 3 阳台 4 流动展览室 5 放映室 6 入口大厅 7 衣帽间
1. exhibition 2. storage 3. balcony 4. room for rotating exhibitions
5. projection room 6. entrance hall 7. wardrobe
A-A' 剖面图 section A-A'

1 展览室 2 入口大厅 3 露台 4 门廊
1.exhibition 2. entrance hall 3. terrace 4. porch
B-B' 剖面图 section B-B'

这个项目将一座原来的马略卡王室住宅（宫殿）改造为帕尔马的公共文化建筑。这座新建筑将临时展览空间和主楼层内的家庭博物馆、报告厅、办公室和一层的餐馆连接到入口的天井处。该建筑的设计目标是将新建筑的开放和公共的特点与原来建筑的家庭特征结合起来，使这两个要素能够和谐共存，并且各具特色。

文化中心的墙体和空间内留有建筑进化的历史遗迹，这些遗迹同时也代表了几个世纪以来马略卡社会的进化过程。墙上累积的时间印记和空间现在的几何形状在新的建筑内展现出来，这主要通过维持房间之间的衔接和光影的密度来实现的，以满足历史悠久的贵族家庭的私人生活。

新元素的加入将建筑物从内到外地进行了改变，使现存空间拥有了新的功能，并在整个宫殿内重组了流线。这一点主要通过让自然光进入到垂直流线内，使穿过中心的漫步道变得更直观且更容易定位来实现的。

建筑师决定将未来并入的结构设置在底层和顶层，中间楼层仍保持了家庭建筑的原始状态。在底层，原有的天井恢复了其半公共的性质，成为一处与街道和宫殿内部相连的过渡空间。展览空间和餐厅围绕着大型天井而建，确保这一楼层的公共性得以实现。而在顶层，破败的木质屋顶结构在经过修复后，形成一座新的博物馆，这座博物馆利用新的采光井可以让自然光进入。这种设计在整个平面中都可见，使不同范

卡萨尔·巴拉格尔文化中心

Flores&Prats + Duch-Pizá

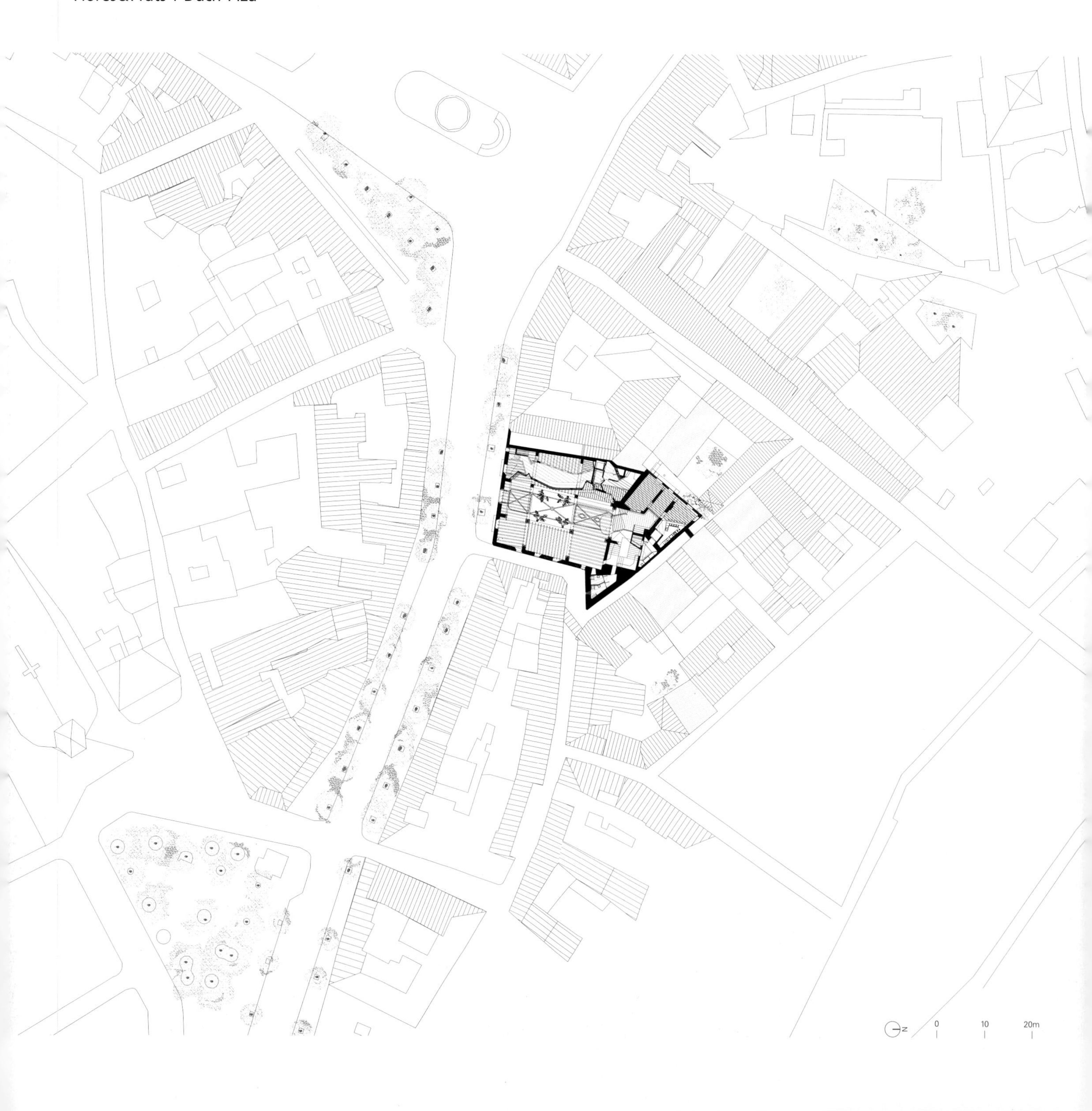

围内的阅读和研究活动都能够围绕着整个庭院和圆屋顶开展。一层之上设有一个报告厅，延伸到露台，占据了屋顶层的空间。这处开放的通风空间的四周是可折叠的屋顶和天窗，也是一条贯穿建筑的人行道的终点，此外，它又一次与一层的入口天井连接起来。

卡萨尔·巴拉格尔文化中心呈现在帕尔马居民的历史和社会记忆当中。它的起源可以追溯到13世纪，延伸至15世纪，一直到18世纪。最后建造的巴洛克式舞台使建筑具备了现代的特征。现在，新的公共空间让宫殿的不同楼层都实现了完全的复兴：包括建造、结构、几何外形和功能方面。将该建筑的所有特色恢复到刚建造时的样子的目的是为了呈现出一种可持续的状态，并且体现一种信念，即遗产需要以一种积极的方式来重新融入到充满生机的城市文化动态当中。

Casal Balaguer Cultural Center

The project converts a former Majorcan royal family house (palace) into a public cultural building for the city of Palma. The new program combines spaces for temporary exhibitions, a house museum on the main floor, a lecture room and offices, and a restaurant on the ground floor linked to the entry patio. The aim of the project is to combine the open, public character of the new program with the domestic qualities of the original building, making possible that these two aspects coexist in parallel, neither dominating the other. The Cultural Center will contain in its walls and spaces the remains of the history of the building's evolution, which at the time represent the phases that the Majorcan society has been through in several centuries. These traces of time accumulated on the walls and the geometry of spaces that arrived until today, are present in the new building by maintaining the articulation among rooms and the density of light and shadows that conformed the private life of this aristocratic family for years.

The intervention transforms the building from inside out, giving new uses to the existing spaces and reorganizing the circulation through the whole palace. This is obtained by linking natural light to vertical circulation, making the promenade through the center more intuitive and easy to orientate.

The main decisions for the incorporation of the future use have taken place on the ground and top floor, being the middle level preserved as the original state of the family house. On the ground floor, the former patio recovers its semipublic character, an in-between space which relates the street with the interior of the palace. Here the exhibition spaces and the restaurant surround the big patio, to ensure a public use of this level. On the top floor, the rehabilitation of the decayed wooden roof structure gave place to a new library, which enjoys the entrance of natural light through new lightwells. This activity occurs in the whole plan, organizing the different ambits of reading and research around the courtyard and the dome. One floor above, a lecture room is extended towards a terrace and occupies the roof level. This open air space, surrounded by the folds of the roof and skylights, ends the promenade through the building and links again with the ground floor entry patio.

The Casal Balaguer is present in the historical and social memory of the inhabitants of Palma. Its origins has been dated in the XIII century, extended in the XV century and then again in the XVIII century. This last Baroque stage is the one that gives the building its present character. The new public occupation performed now, supposes a complete rehabilitation of the palace at different levels: constructive, structural, geometrical and of uses. This aim to recuperate all the qualities of the found state represents a sustainable statement, with the conviction that heritage needs to be reincorporated in an active way in the living cultural dynamics of the city.

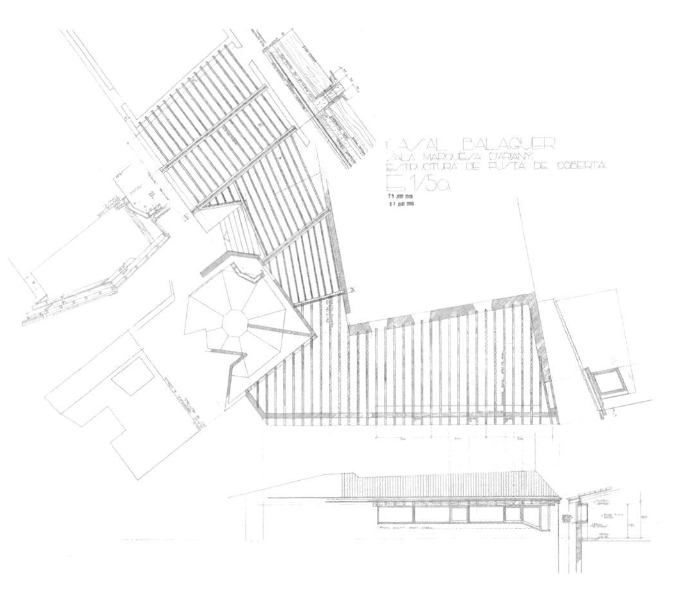

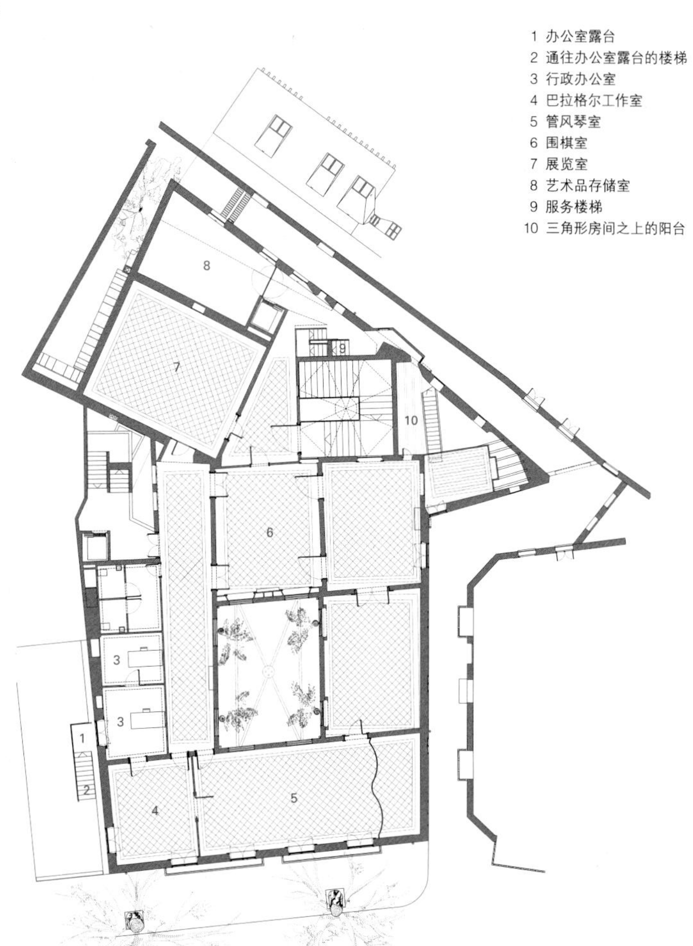

1. offices' terrace 2. stairs to offices' terrace 3. administration offices 4. Balaguer's studio 5. organ room 6. chess room 7. exhibition room 8. storage of artworks 9. service stairs 10. balcony over triangular room

二层 second floor

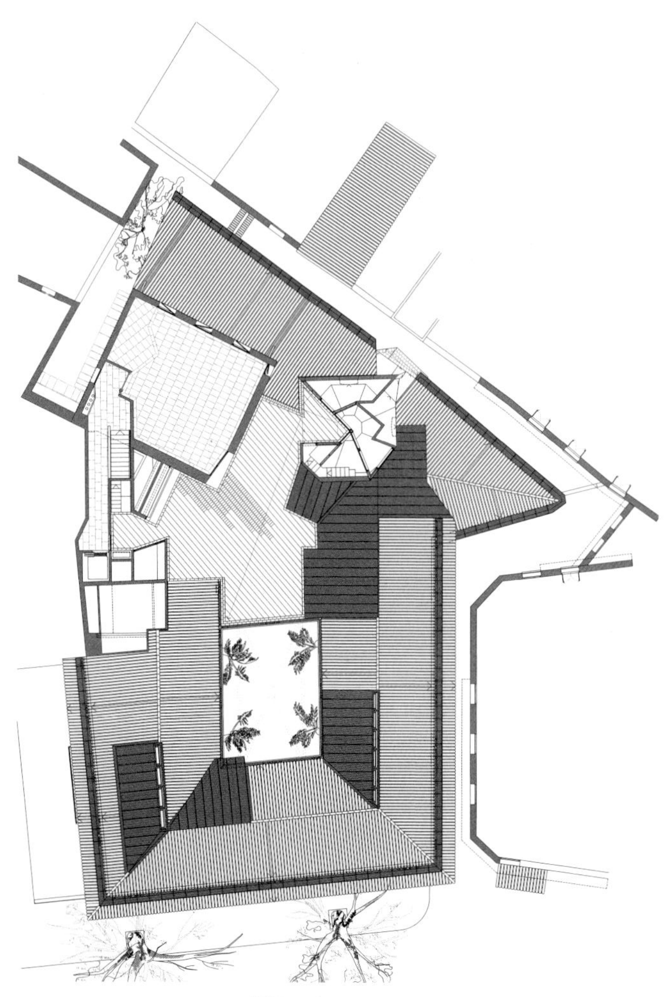

屋顶 roof

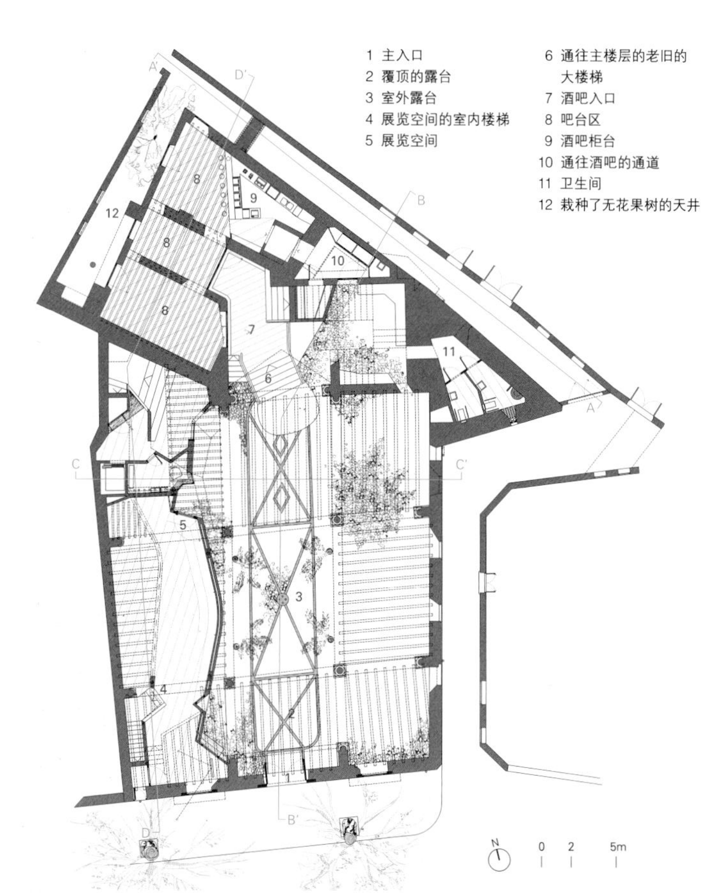

1. main access 2. covered patio 3. outdoor patio 4. interior staircase of exhibition space 5. exhibition space 6. old main staircase to the main floor 7. bar entry 8. area for bar tables 9. bar counter 10. access to bar 11. WC 12. fig tree patio

一层 first floor

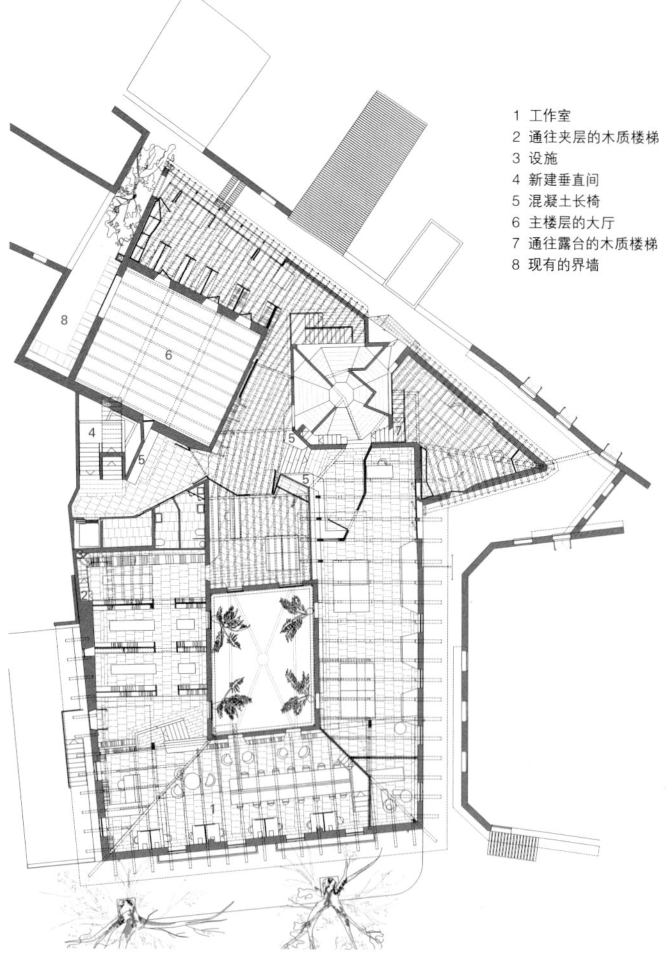

1. workshop 2. wooden staircase to mezzanine 3. installations' shaft 4. new vertical core 5. concrete bench 6. hall on the main floor 7. wooden staircase to terrace 8. existing party wall

三层 third floor

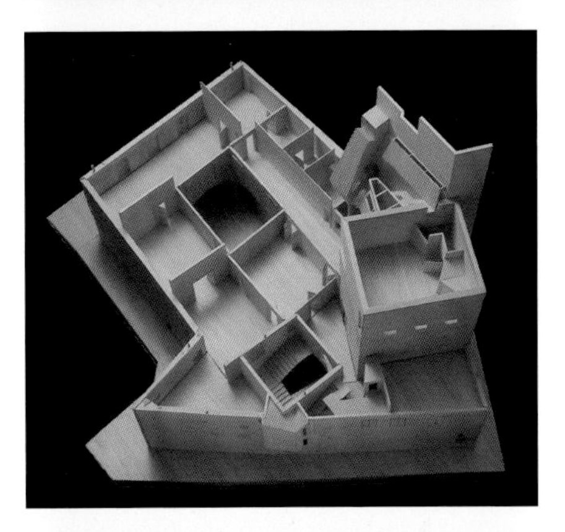

项目名称：Casal Balaguer Cultural Center
地点：Calle Unión 3, Palma de Mallorca
建筑师：Flores & Prats, Barcelona + Duch-Pizá, Palma
合作者：Caterina Anastasia, Ankur Jain, Els van Meerbeek, Cristian Zanoni, Carlos Bedoya, Guido Fiszson, Ellen Halupczok, Julia Taubinguer, Paula Ávila, Nicolás Chara, Eugenia Troncoso, Israel Hernando, Hernán Barbalace, Benedikte Mikkelsen, Mar Garrido, Celia Carroll, Jorge Casajús, Juan Membrive, Oriol Valls, Tanja Dietsch, Sergi Madrid, Sergio Muiños, Lucas Wilson, Anna Reidy, Maria Amat Busquets, Fabrizia Cortellini, Veronica Baroni, Elvire Thouvenot, Carlotta Bonura, Francesca Tassi-Carboni, Tomás Kenny
发起人：Palma City Council / 考古专家：Grupo Arqueotaller
历史研究与记录：Ma. Dolores Ladaria
材料分析：Lend Consulting / 结构设计：Fernando Purroy
功能：complete rehabilitation of a Baroque palace of palma as a cultural center, house museum and library specialized in art
用地面积：870m² / 总建筑面积：2,300m² / 有效楼层面积：2,500m²
设计时间：1996_first partial project for the roofs / 2001—2003 _ second global project for the building
施工时间：2009—2010_first phase / 2011—2013_second phase / 2014_third phase under construction
竣工时间：2014 / 摄影师：©Adrià Goula (courtesy of the architect)

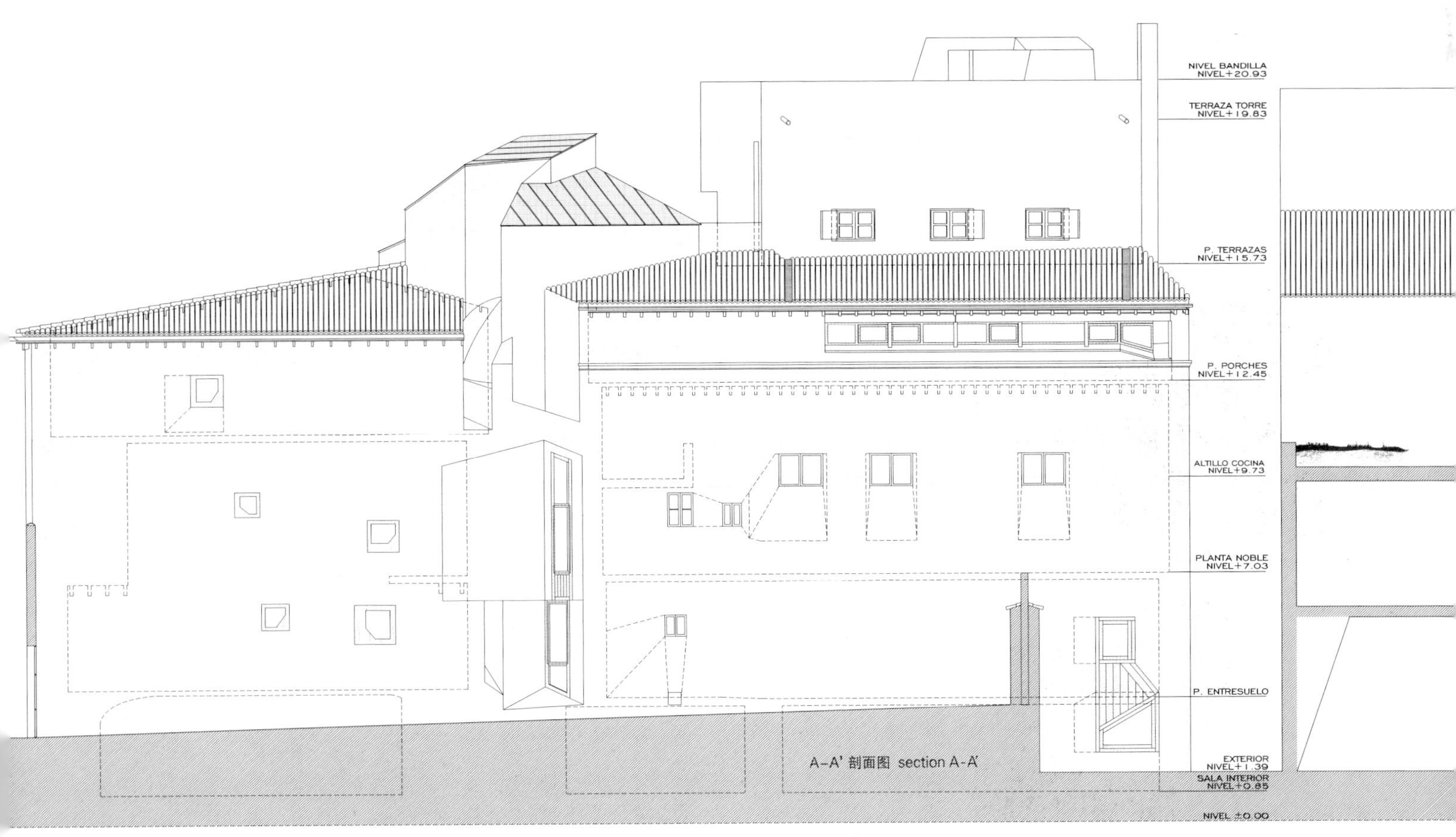

A–A' 剖面图 section A-A'

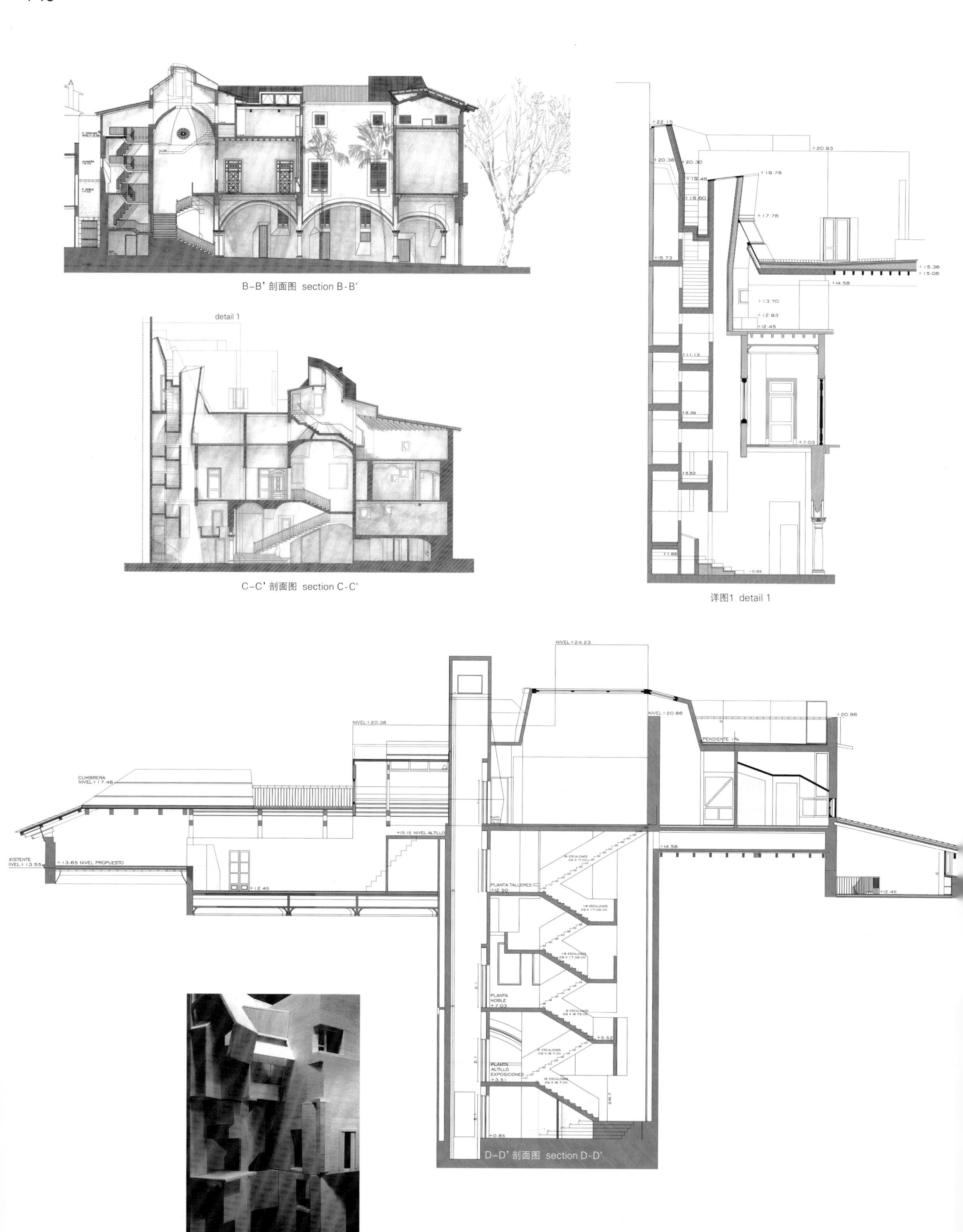

B-B' 剖面图 section B-B'

C-C' 剖面图 section C-C'

详图1 detail 1

D-D' 剖面图 section D-D'

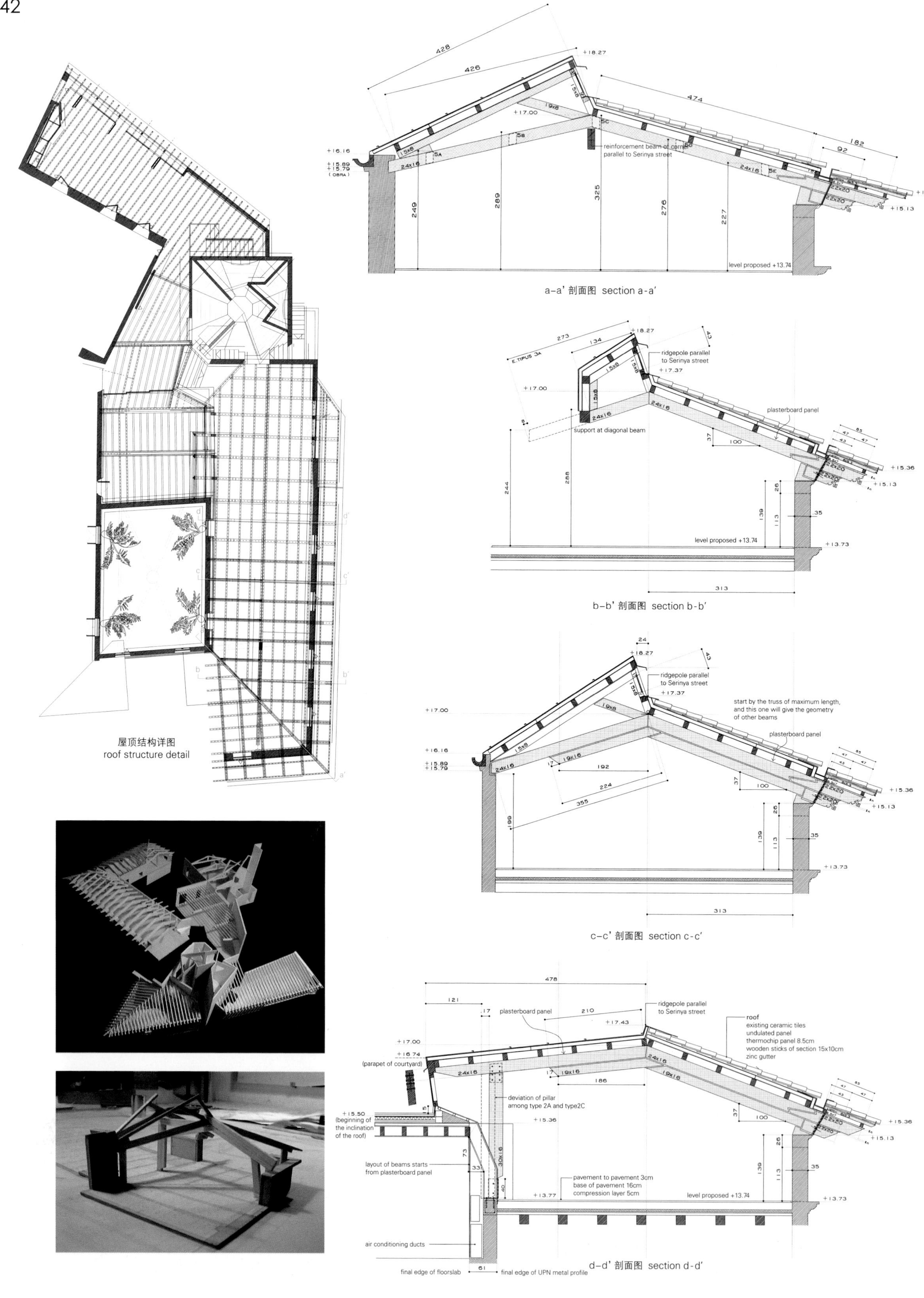

屋顶结构详图
roof structure detail

a-a' 剖面图 section a-a'

b-b' 剖面图 section b-b'

c-c' 剖面图 section c-c'

d-d' 剖面图 section d-d'

AULAS DE
DIBUJO Y
PINTURA
ALREDEDOR
DE LA
CÚPULA EN EL
NIVEL DE BAJO CUBIERTA
PALAU
BALAGUER
PALMA DE MALLORCA
OCTUBRE 2003
ESCALA GRÁFICA

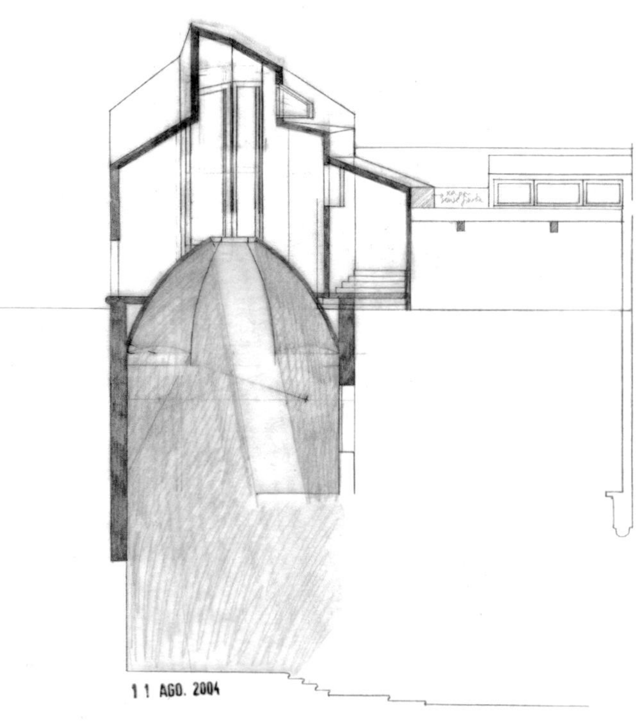
1 1 AGO. 2004

哈塞尔特大学

noAarchitecten

在更名为哈塞尔特大学后，该学校也渴望在城市中心获取更明显的存在感。一个法学院的建立使这一愿望成真，而一处监狱旧址成为合适的场地。但轻微的恐慌随之而来：小规模的、开放型的大学真地能进入一座象征权力工具的大楼吗？

这个城市校园由三栋建筑构成。监狱旧址是公众的焦点，用于举办讲座和研讨会的城市接待区。法学院位于其后，校长的办公室位于隔壁，沿城市环路设置。

监狱于1859年建成，虽然它看起来像是一座星状圆形监狱，其实不然。从宏伟的中央大厅望去，你可以看到牢房的末端墙体，与外环走廊背靠背连接。大厅里没有中央控制站，但有一个讲坛，以期宗教和个体监禁的结合能够引导囚犯们看清自己犯错的方式。

如果你看得够久，你会发现，正是最初那些引起人们厌恶的特征提供了附加价值。监狱的墙体变身力量之源，成为城市内举世闻名的地标，满载深意。墙体似乎有了新的内涵。困在监狱里的这种“消磨时间的权力”取代了监禁生活的形象。多条横向通道使两座礼堂和一个自助餐厅相互映衬，在楼翼之间延伸到室外区域。监狱迷宫般的自然特性为在监狱里建造另一番天地提供了意想不到的可能性。这座建筑也因此成为

一个小镇，有几个出入口、广场、街道、庭院及出人意料的屋顶花园，要知道，这座墙体以前最多不过是矮墙罢了。

教职工楼坐落在校园后方的土地上，楼里有学生教室和教师办公室。在这里，建筑规模非常重要，向上倾斜的扇形区域缓冲了大型结构和自行车道一侧的住宅之间的冲突。这栋建筑非常引人注目，有着宽阔敞亮的走廊、开放的木楼梯和充足的空间供学生使用。红砖、木门和回转窗都是传统学校建筑的标志元素。建筑有宽敞的入口，带有一个独特的混凝土雨篷、一个时尚的长凳以及一个三岔灯柱。此外还有一条地下通道通到监狱建筑，且与一座户外礼堂在空间上相互交织，产生一处令人意想不到的、复杂且明亮的空间。

第三座楼非常具有代表性，是城市环形路的一个弯角，高耸且紧凑。你可以开车从其底部穿行，到达监狱下方的新停车场。立面覆有几层绿色玻璃，在其重叠的地方形成了深绿色区域。此刻，在我们面前，校长办公室与监狱旧址，绿色与红色，透明与封闭，两两相互抗衡着。而在里面，水磨石、木薄片、玻璃和镜子共同创造了一处精致的空间。中央楼梯决定了每一层楼的空间排列，就像一个万花筒一样。

Hasselt University

After the name of the institution was changed to Hasselt University, there was also a desire to acquire a more pronounced presence in the city center. When a Faculty of Law was established, this desire took concrete form. The search for a suitable location led to the former prison. Slight panic ensued. Can a small-scale, open university really move into a building conceived as an instrument of power?

This urban campus consists of three buildings. The former prison is the public focal point, and also takes on the role of urban reception area for lectures and symposium. The Faculty of Law stands on the land behind it and the Rector's Office is sited next door along the city's ring-road.

The prison dates from 1859. Although it looks like a star-shaped panopticon, it isn't. From the monumental central hall you see the end walls of the cells, built back to back with the corridors running along the outside. The hall did not originally contain a central control post, but a pulpit. It was intended that a combination of religion and individual isolation would lead the prisoners to see the error of their ways.

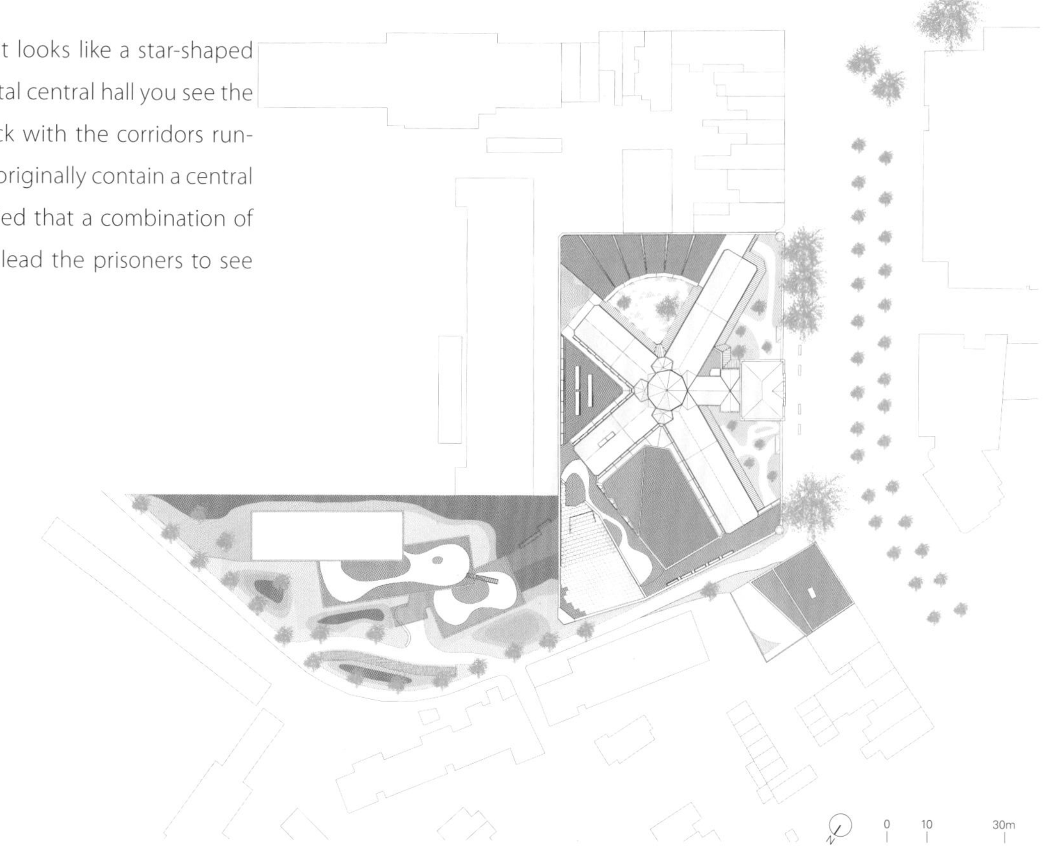

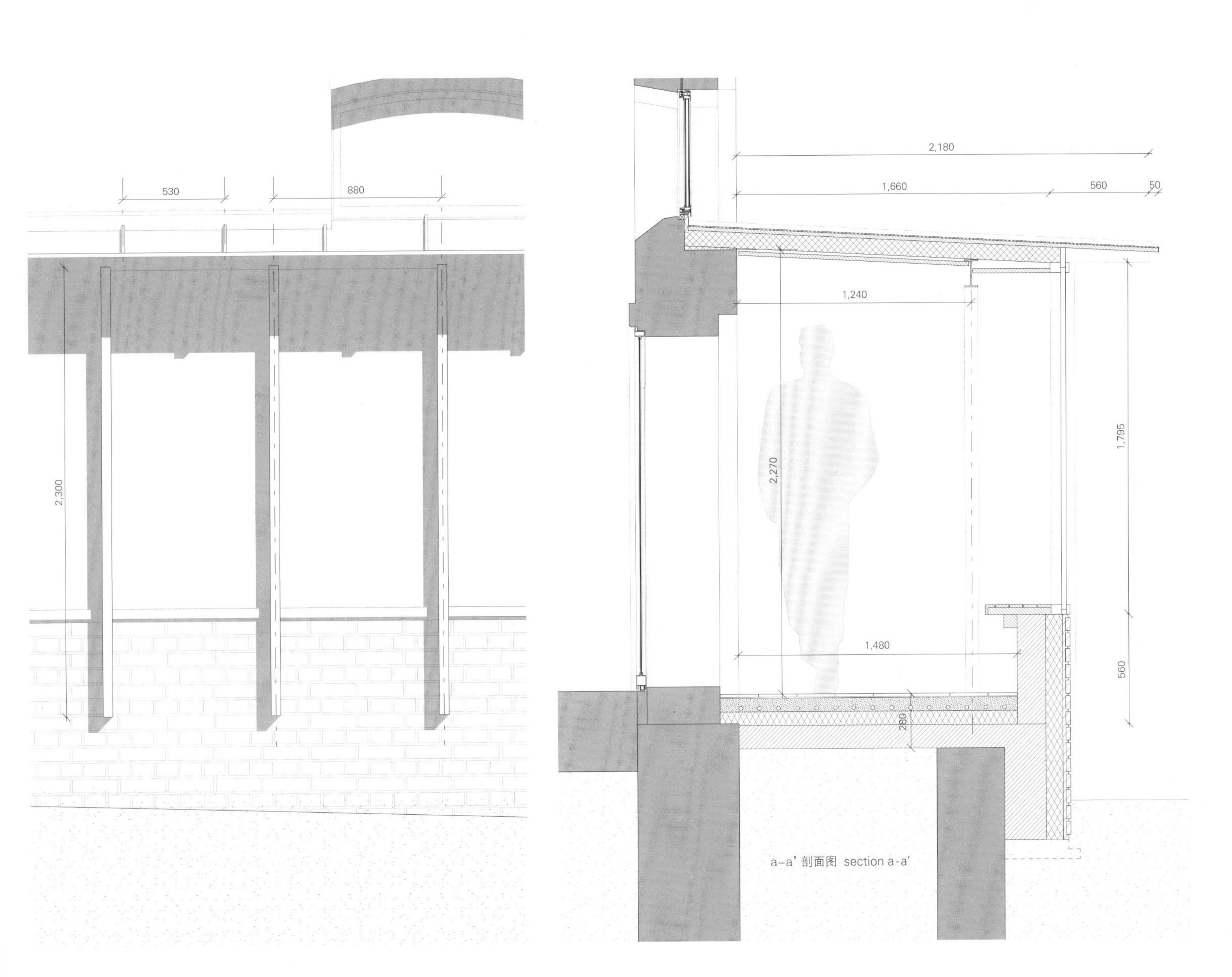

a–a' 剖面图 section a-a'

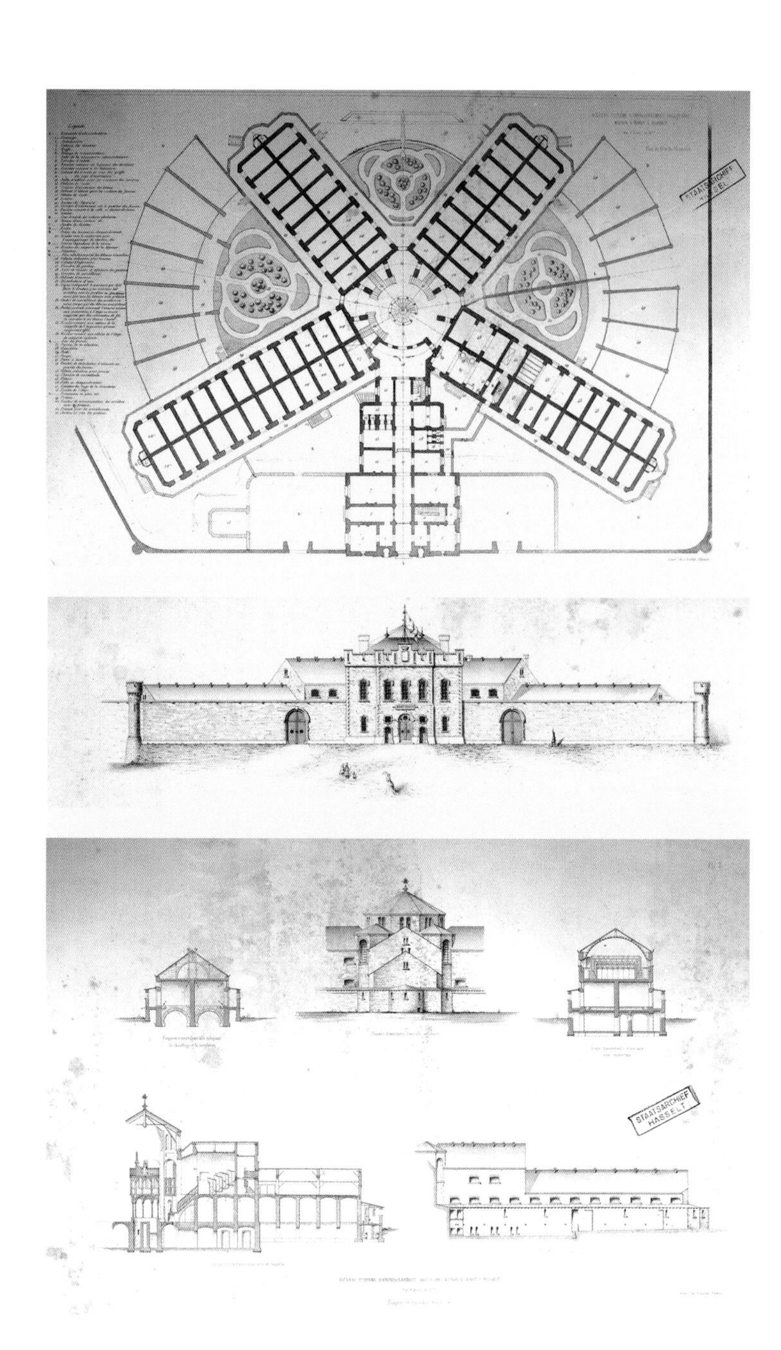
STAATSARCHIEF
HASSELT

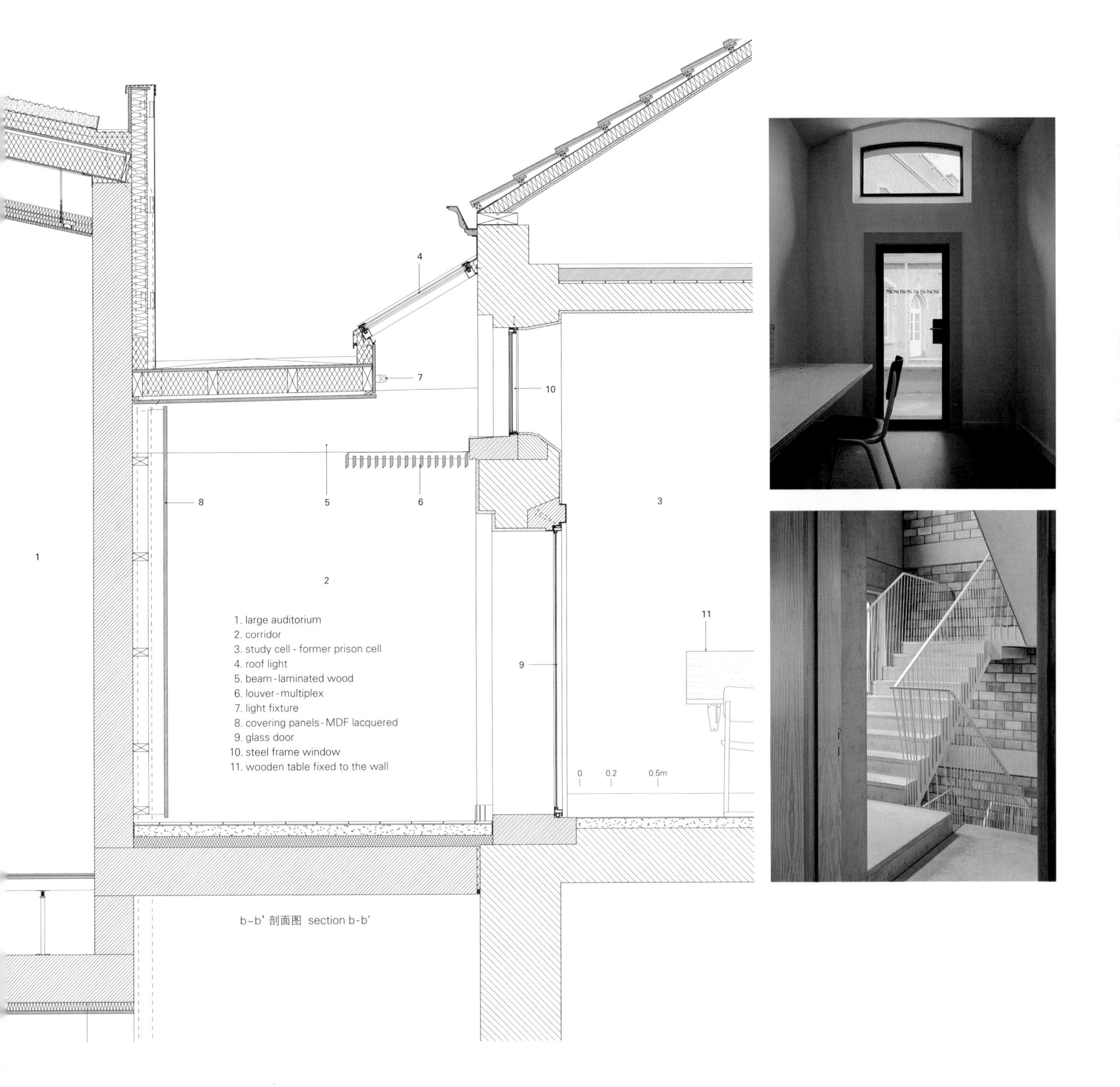

b–b' 剖面图 section b-b'

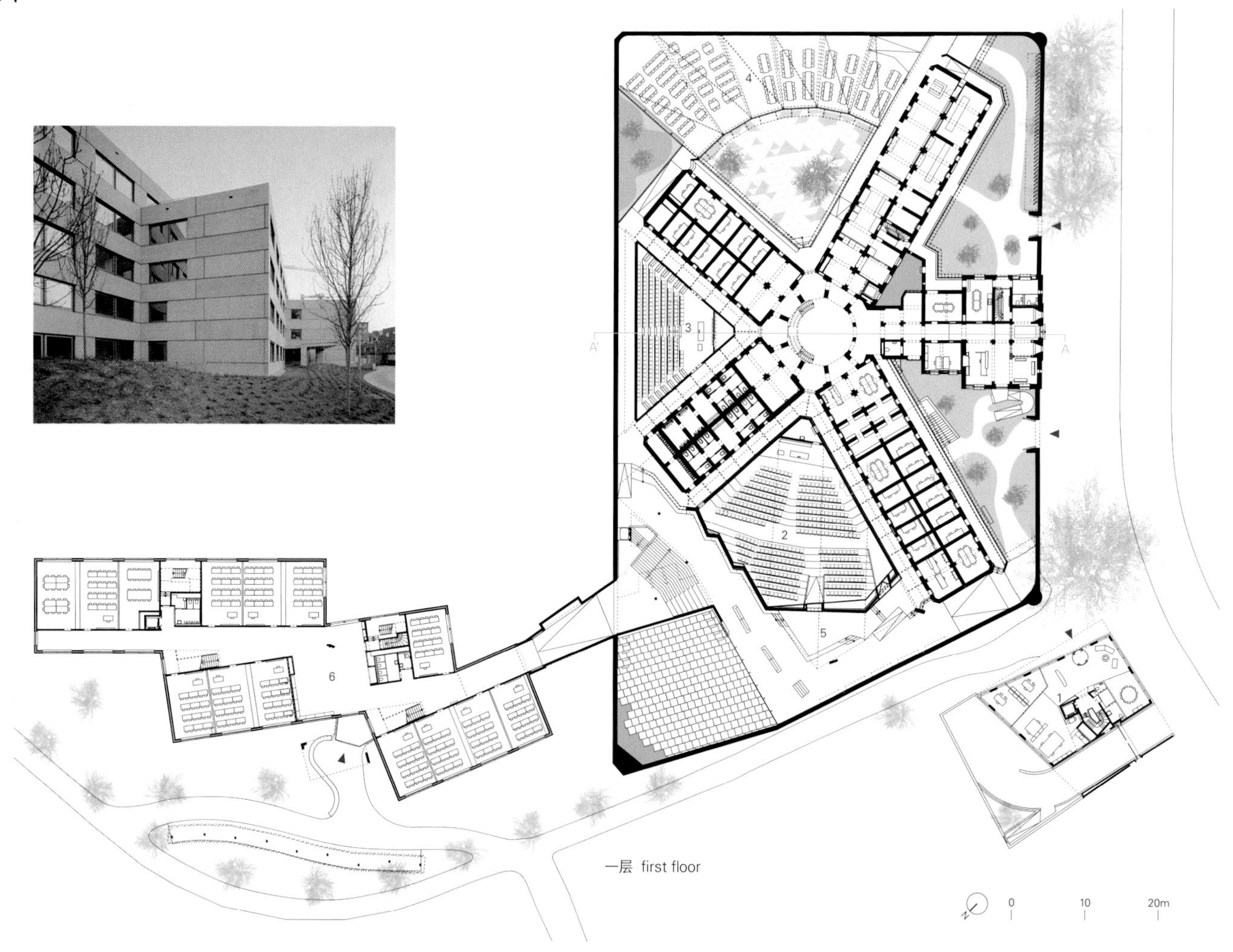

一层 first floor

项目名称：Hasselt University
地点：Martelarenlaan 42, 3500 Hasselt, Belgium
建筑师：noAarchitecten
项目团队：Leen Corthier, Danny Dezutter, An Fonteyne, Greg Geertsen, Sander Laureys, Beatrijs Noppe, Christiaan Oomen, Kim Pecheur, Marijn Proot, Jitse van den Berg, Arnout Van Vaerenbergh, Pieter Verreycken, Peter Verstraete, Philippe Viérin, Tim Wallyn
结构工程师：studieburo Mouton
服务工程师：RCR
景观建筑师：Jan Minne
音效工程师：Daidalos Peutz
整体艺术设计师：Philip Van Isacker, Benoît van Innis
用地面积：9,400m² / 总建筑面积：6,300m²
有效楼层面积：15,900m²
设计时间：2008 / 竣工时间：2013
摄影师：©Kim Zwarts

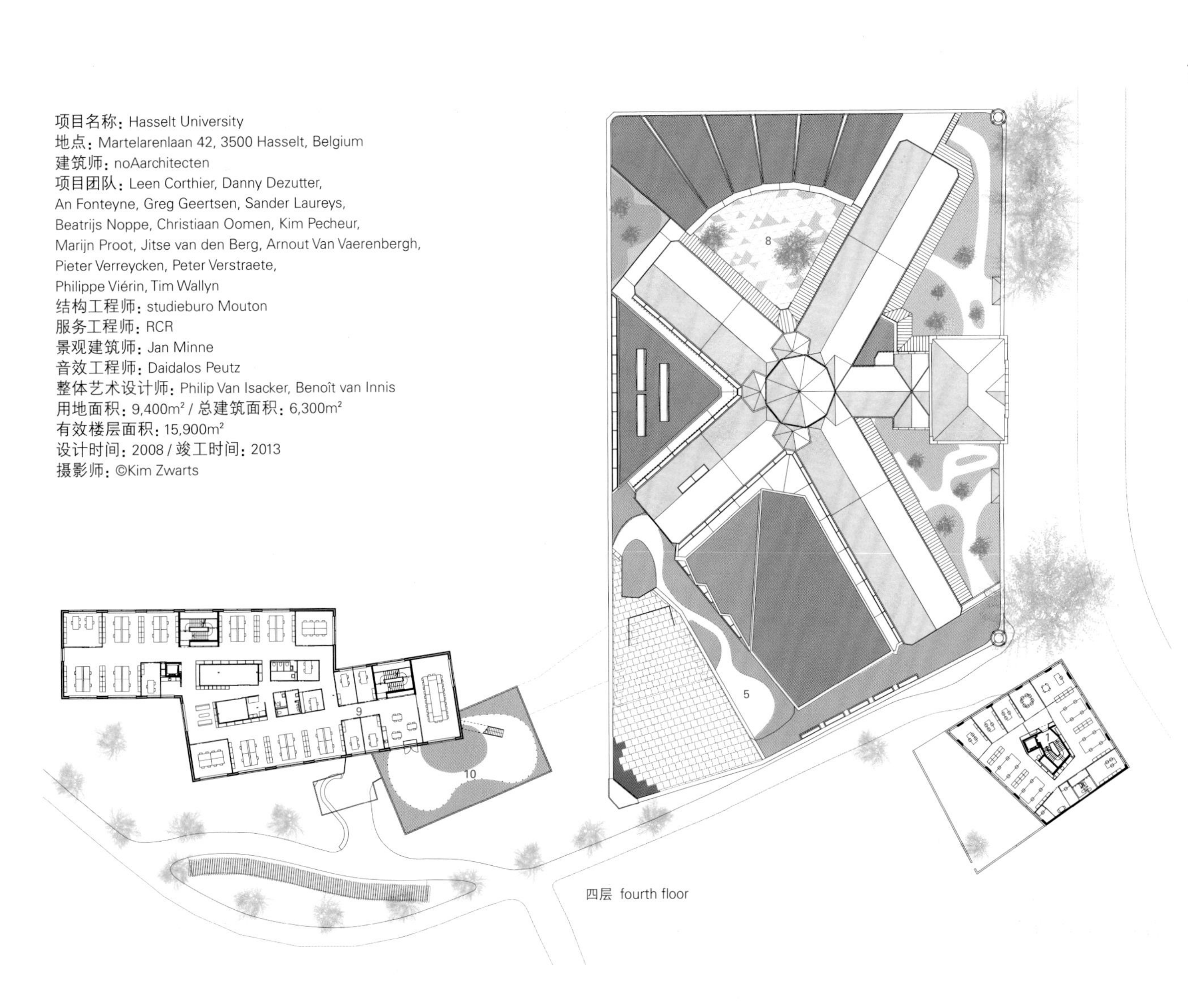

四层 fourth floor

1 校长办公楼
2 大型礼堂
3 小型礼堂
4 自助餐厅
5 集合区
6 教学楼——教室
7 集合区——屋顶花园
8 自助餐厅花园
9 教学楼——办公室
10 屋顶花园

1. rector's office building
2. large auditorium
3. small auditorium
4. cafeteria
5. agora
6. faculty building–classrooms
7. agora–roof garden
8. cafeteria garden
9. faculty building–offices
10. roof garden

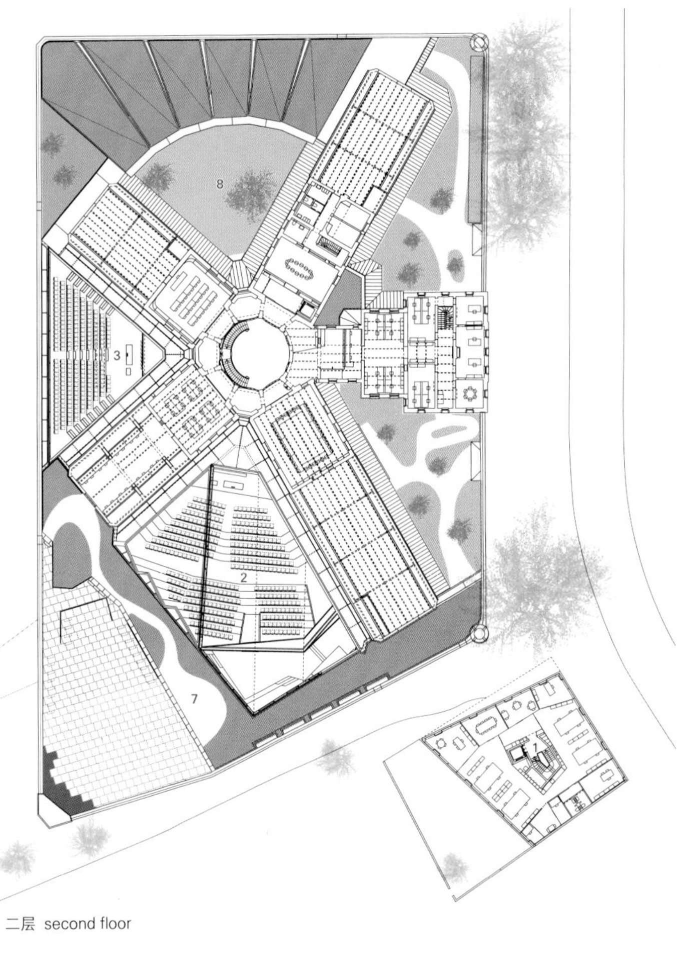

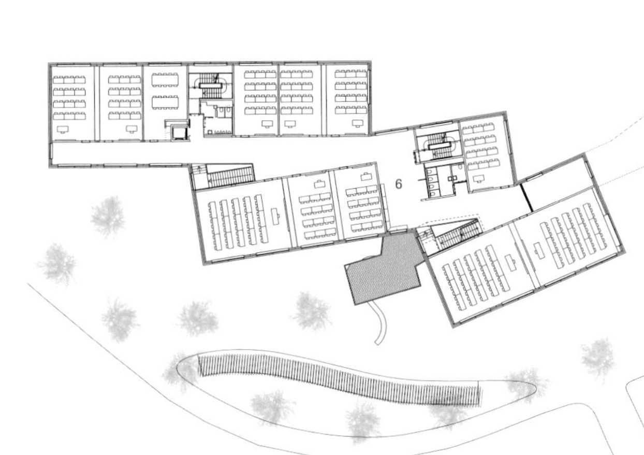

二层 second floor

If you look long enough, it is those characteristics that initially arouse aversions that turn out to offer added value. The prison wall forms the strength of the place, being a universally-known landmark in the city and a bearer of much meaning. The wall seemed to have generated new connotations. The privilege of spending time inside the wall replaces the image of imprisonment. The presence of so many lateral corridors made it possible to fit two auditoriums and a cafeteria into the outdoor areas between the wings. Its labyrinthic nature offered unexpected potential for the creation of a separate world inside the prison wall. The building thus becomes a small town, with several entrances and exits, squares, streets, courtyards and unexpected roof gardens for which the prison wall acts as no more than a parapet.

The faculty building with its classrooms and lecturers' rooms nestles into a piece of land at the back. Scale is important here, and the upward-sloping fan shape eases the confrontation between the large volume and the terraced houses on the other side of the cycle path. It has become an inviting building with broad, light corridors, open wooden staircases and a surfeit of space where students can spend their time. The red bricks, wooden doors and pivot windows are elements from traditional school buildings. The building has its own spacious entrance with a distinctive concrete canopy, a stylish bench and a three-branched lamp-post. It has a subterranean passageway to the prison building and it is spatially interwoven with an outdoor auditorium, which makes for an unexpectedly complex and light space.

The third building has a representative role. It is a corner-piece on the city's ring-road. It is tall and compact. You can drive under it to a new car park beneath the prison. The facade consists of several layers of green glass, with dark green zones created where they overlap. So what we have here is Rector's Office versus former prison, green versus red, transparent versus closed. Inside, a refined space is created using terrazzo, wood veneer, glass and mirrors. The central stairwell determines the spatial arrangement of each floor and acts like a kaleidoscope.

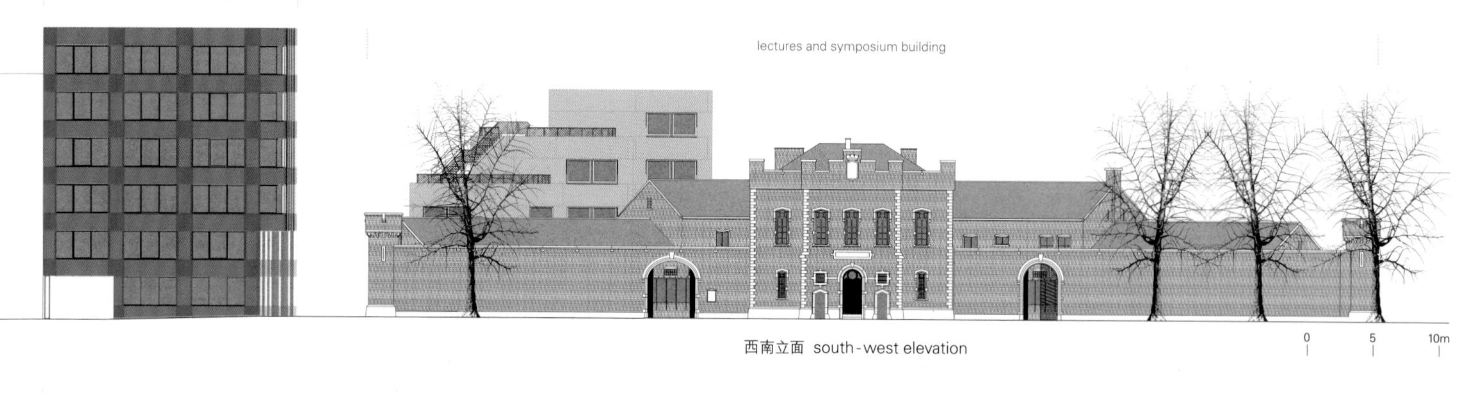

西南立面 south-west elevation

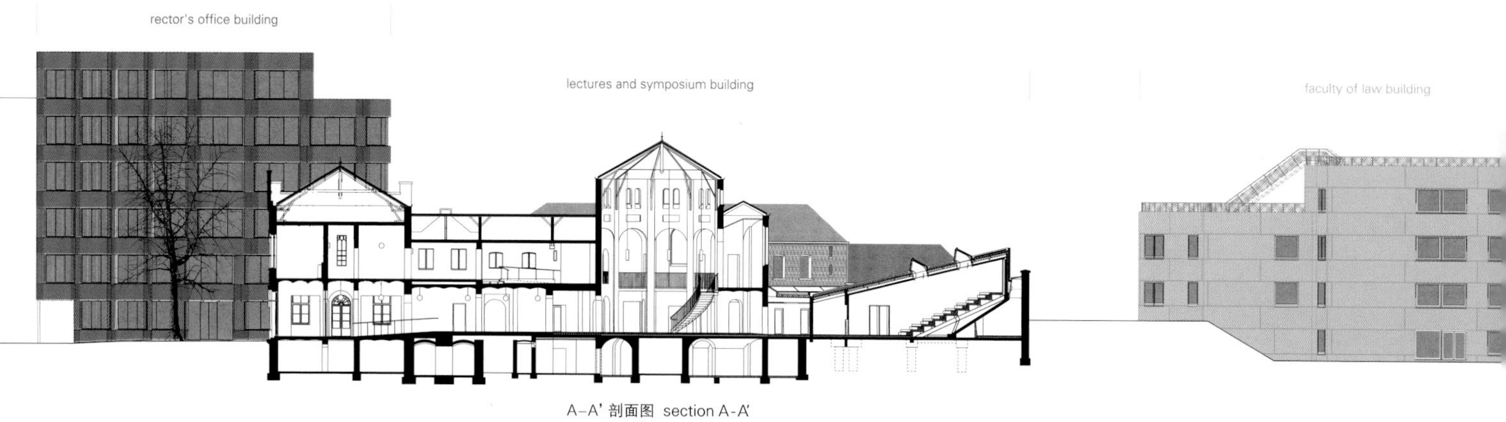

A–A' 剖面图 section A-A'

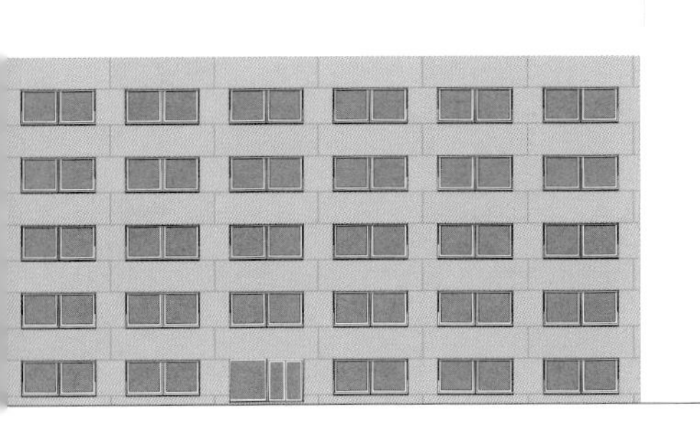

哈佛艺术博物馆

Renzo Piano Building Workshop

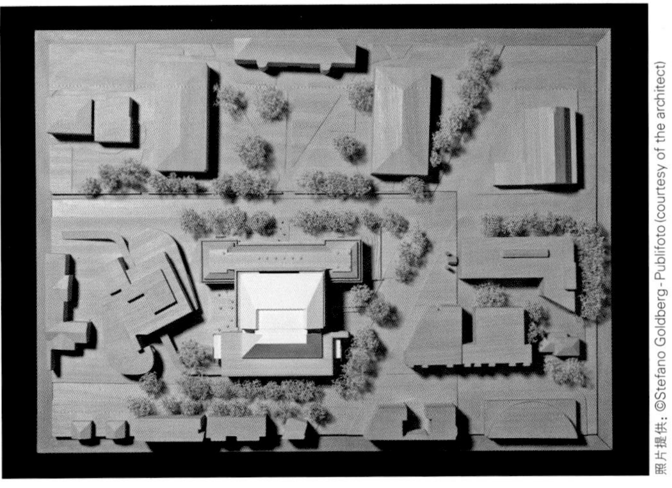

照片提供：©Stefano Goldberg-Publifoto (courtesy of the architect)

哈佛大学的三座艺术博物馆——Fogg艺术博物馆、Busch-Reisinger博物馆和Sackler博物馆——合并成一个重新规划的、升级的设施，即哈佛艺术博物馆，位于昆西街的福格艺术博物馆的现有场地。福格艺术博物馆的修复的古老庭院位于面积为18500m²的新博物馆的核心位置。

新建筑将福格艺术博物馆的被保留的、于20世纪20年代乔治亚时代建造的修复建筑与东部的一座沿普雷斯科特街建造的增建结构结合起来。同时一个崭新的玻璃屋顶结构将旧建筑和新建筑结合起来。这个屋顶增建结构的设计充分考虑了周围历史结构的敏感性，允许有限的自然光进入安全实验室、学习中心、画廊和下方的庭院内。

于1920年建造的，由库里奇、谢普利、布尔芬奇和阿伯特建筑师事务所设计的原有建筑，是第一座将博物馆空间、教学空间和保护空间融为一体的、促进学术成就的设施。新中心也遵循了这一传统，使200 000件收藏品更加便于教学和学习使用。

所有1925年后的增建建筑和改建建筑都被拆除了，来为普雷斯科特街的新扩建建筑留出空间。历史建筑的各个方面——结构、机械和科技方面，都会得到修缮和升级。

画廊和学习中心进行了大规模的扩建；学习中心为了适合博物馆功能方面的重要性，位于四楼，即建筑的中心位置。检测实验室仍然位于建筑的顶部，即学习中心之上，新建的斜坡式玻璃屋顶之下。公共设施和举办特殊活动的援助空间都进行了扩建以及现代化改造，包括在地下室层建造一座设有294个座位的观众席。

博物馆最初的入口面向大学校园，现在，普雷斯科特街一侧设置的新入口使博物馆象征性地向社区开放。从内部庭院直到入口的建筑物两侧的视野可以帮助访问者定位自己的方向，同时，这里还会有其他的视野，穿过咖啡馆、商店到达百老汇和邻近的卡朋特艺术中心。

一座冬季花园从主画廊的外部突出，位于扩建结构的北端。这座花园和一楼展览空间立面的其他玻璃部分能使人们从街上就看到博物馆，并渐进地让日光射入建筑物当中。

照片提供：©Aerial by Lesvants.com (courtesy of the architect)

1 入口游廊　3 画廊　5 博物馆商店　7 男士卫生间
2 走廊　4 庭院　6 女士卫生间

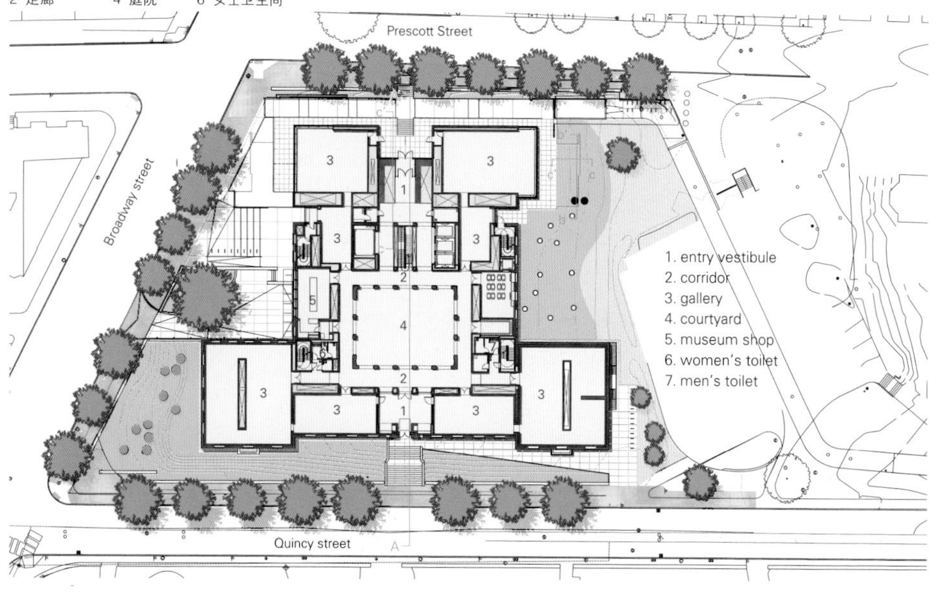

项目名称：Harvard Art Museums
地点：Cambridge, Massachusetts, USA
建筑师：Renzo Piano Building Workshop
合作商：Payette
主要合作者：M.Carroll, E.Trezzani
合作者：J.Lee, E.Baglietto, S.Ishida
CAD绘图师：R. Aeck, F. Becchi, B. Cook, M. Orlandi, J. Pejkovic, A.Stern and J. Cook, M. Fleming, J. M.Palacio, S. Joubert; M. Ottonello
模型制作：F. Cappellini, F. Terranova, I. Corsaro
结构工程师：Robert Silman Associates
MEP工程师 / 照明设计师 / 立面工程师 / Code&LEED顾问：Arup
土木工程师：Nitsch Engineering
木材专家：Anthony Associates
成本顾问：Davis Langdon
音效工程师：Sandy Brown Associates
树艺家：Carl Cathcart
修复顾问：Building Conservation Associates
甲方：Harvard Art Museums
用地面积：6,750m²
总建筑面积：2,800m²
有效楼层面积：18,952m²
设计时间：2006 / 竣工时间：2014
摄影师：©Nic Lehoux (courtesy of the architect)(except as noted)

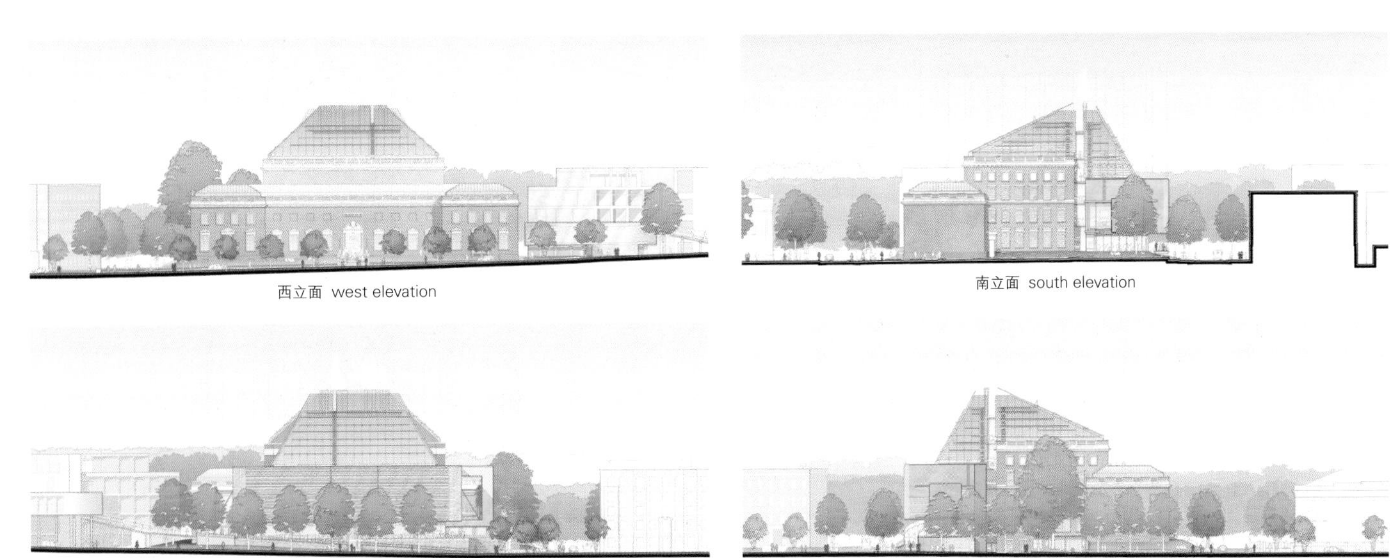

西立面 west elevation

南立面 south elevation

东立面 east elevation

北立面 north elevation

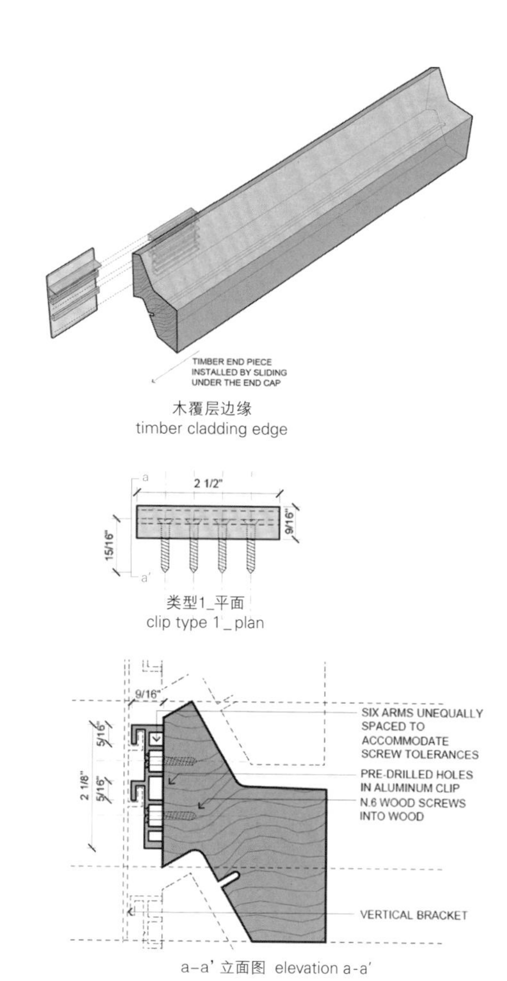

木覆层边缘
timber cladding edge

类型1_平面
clip type 1_plan

a–a' 立面图 elevation a-a'

1. painted steel cross bracing
2. interior timber cladding
3. stainless steel cables
4. unfaced mineral wool insulation
5. 1/4" aluminum panel
6. intermediate transom
7. 10x25 C-channel
8. exoskeleton vertical panel
9. sliding timber panel

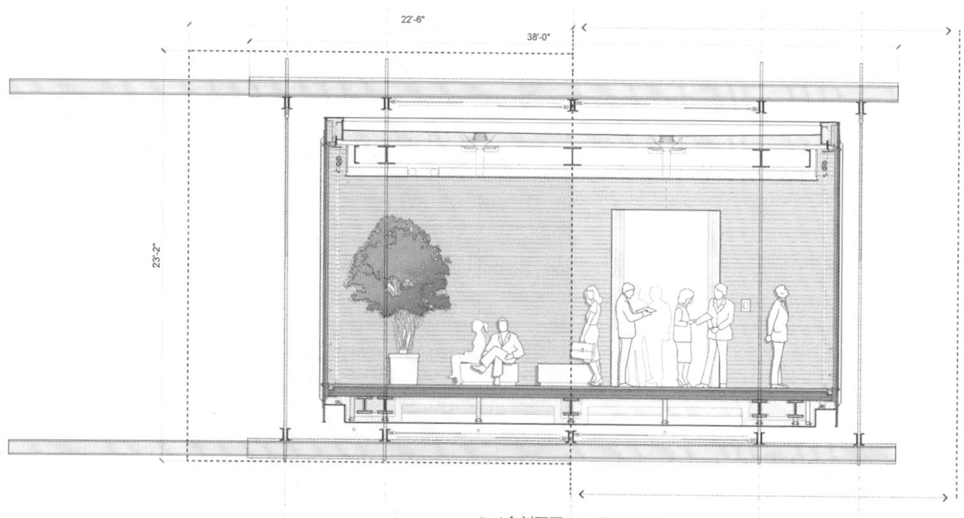

b–b' 剖面图 section b-b'

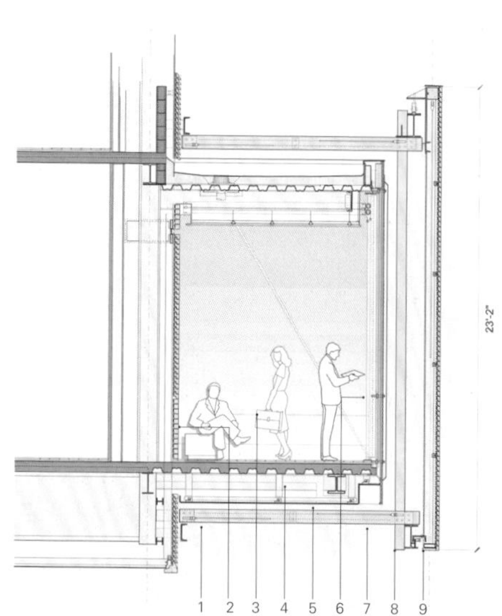

c–c' 剖面图 section c-c'

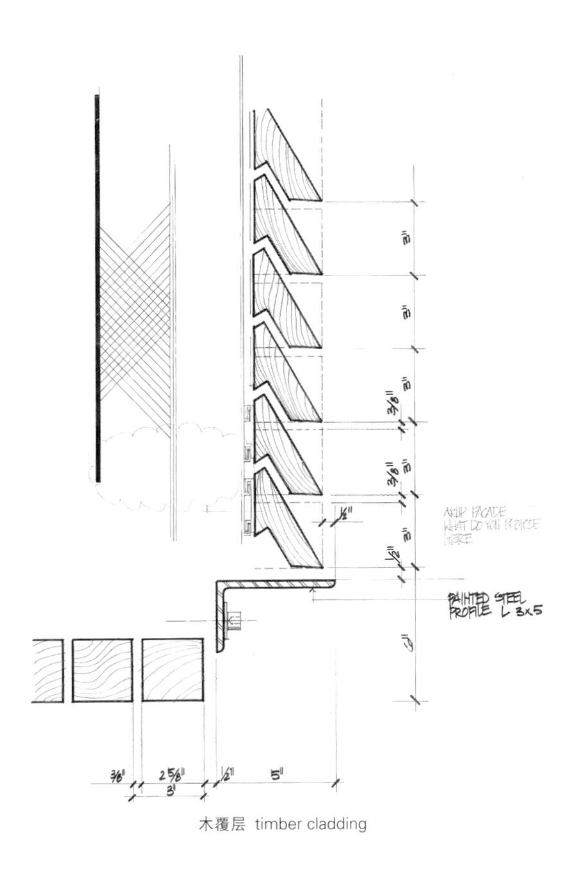

木覆层 timber cladding

Harvard Art Museums

Harvard University's three art museums – the Fogg, the Busch-Reisinger and the Sackler – are being consolidated into one reorganized and upgraded facility, Harvard Art Museums, on the current site of the Fogg Museum on Quincy Street. The restored historic courtyard of the Fogg Art Museum will be at the heart of 200,00 sq. ft (18,500 sq.m) of new museum space.

The new facility will combine the Fogg's protected 1920's Georgian revival building, with a new addition on its east side, along Prescott Street. A new glazed rooftop structure bridges the old and the new. The rooftop addition, designed with sensitivity to surrounding historic structures, will allow controlled natural light into the conservation lab, study centers, and galleries, as well as the courtyard below.

The original 1920's building by Coolidge, Shepley, Bulfinch and Abbot Architects, was the first of its kind, combining museum space, teaching and conservation in one facility to promote scholarship. Following this tradition, the new centre is designed to make the collection of 200,000 objects more accessible for teaching and learning.

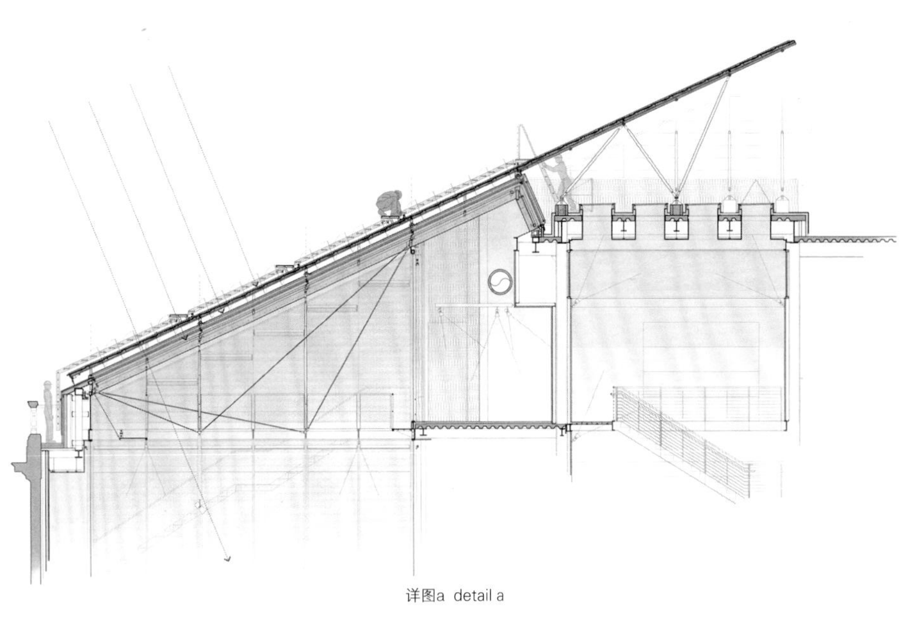

详图a detail a

照片提供：©Michel Denancé (courtesy of the architect)

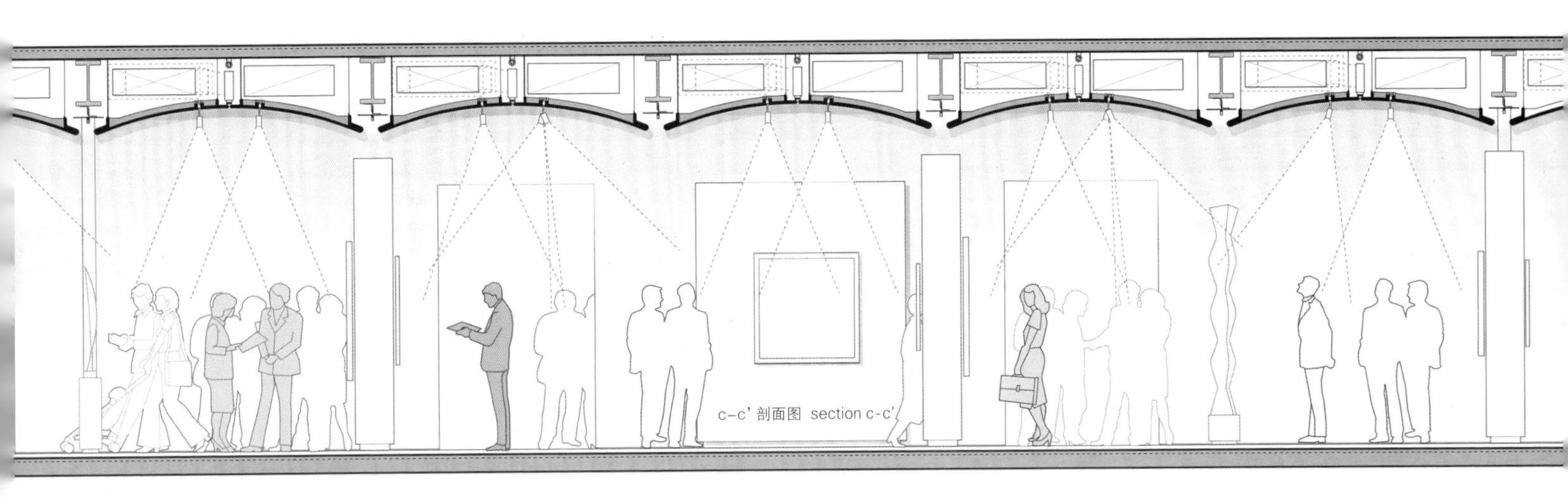
c–c' 剖面图 section c-c'

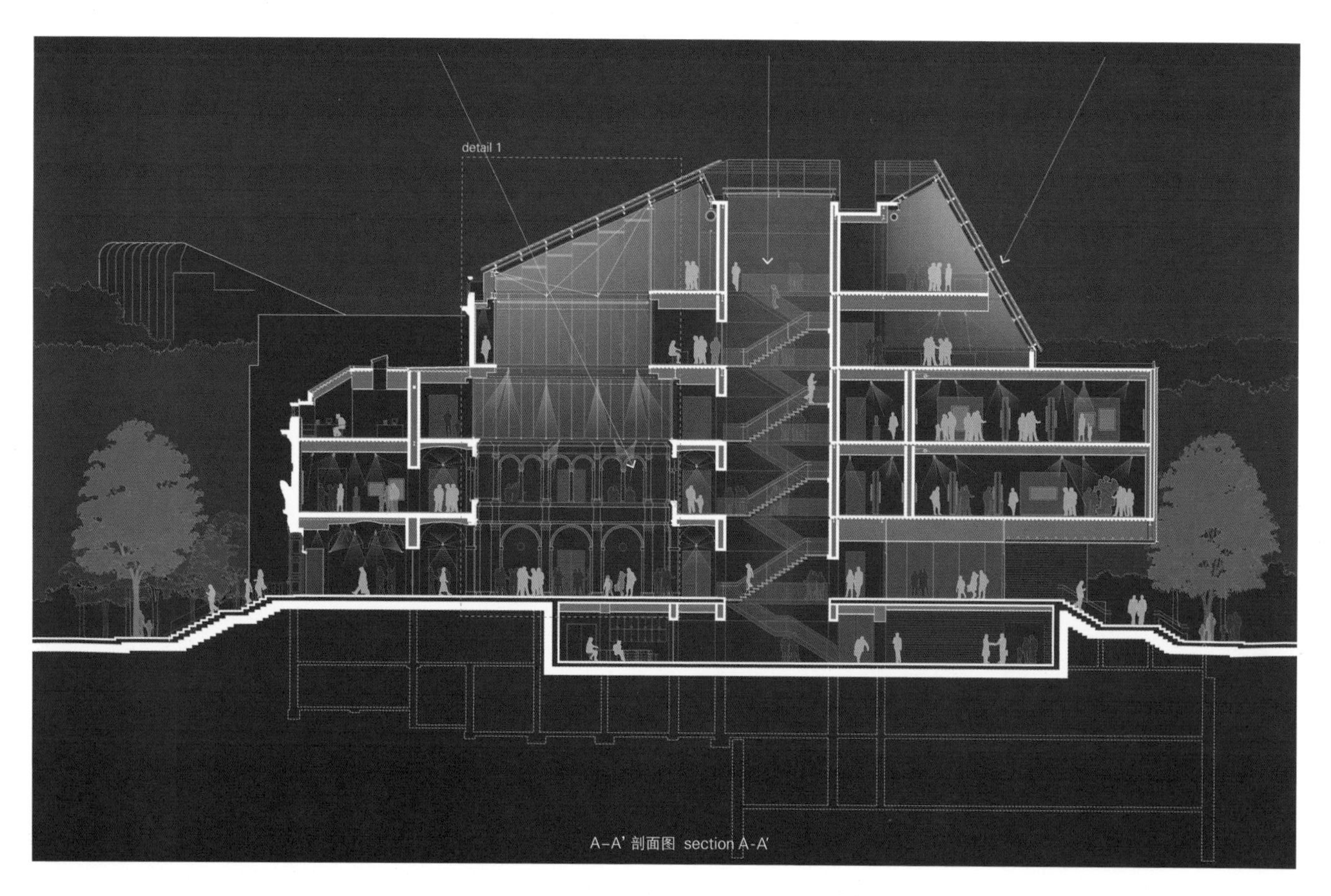

A–A' 剖面图 section A-A'

All post-1925 additions and alterations have been demolished to make way for the new extension on Prescott Street. All aspects of the historic building – structural, mechanical and technical – will be restored and upgraded.

Galleries and study centers are being significantly expanded; as befits their importance to the mission of the museum, the study centers are at the center of the building on level four. The conservation lab will continue to occupy the top of the building, above the study centers under the new sloping glazed roof. Public amenities, and support spaces for special events will be enlarged and modernized, and include an auditorium of 294 seats at basement level.

While the original entrance faces onto the university campus, a new entrance into the museum from Prescott Street symbolically opens the museum to the local community. Views from the interior courtyard through to the entrances on both sides of the building will help visitors to orientate themselves and there will also be secondary views, through the cafe and the shop, to Broadway and the Carpenter Center next door.

At the north end of the extension a winter garden projects beyond the main gallery volume. This and other glazed sections of facade in the first-floor exhibition space allow views into the museum from the street and bring daylight into the building in a very controlled way.

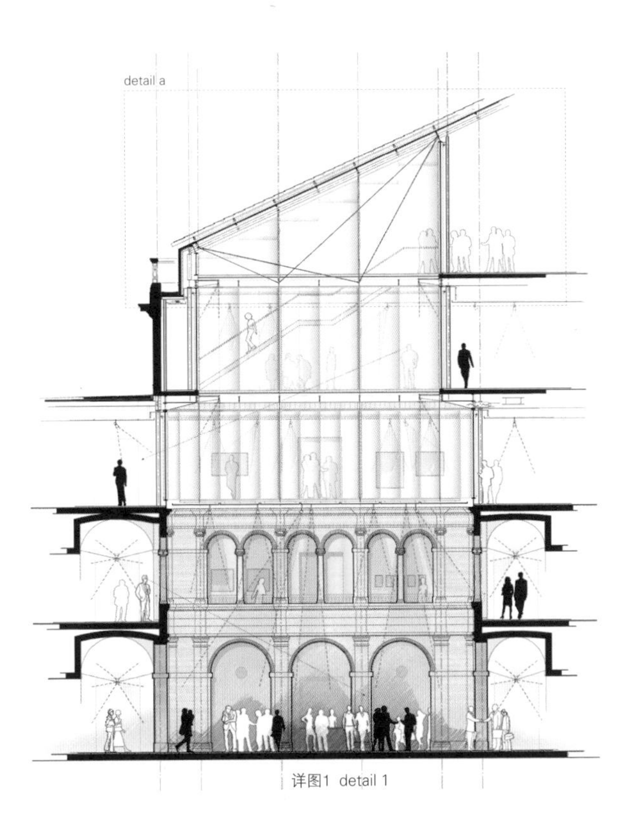

详图1 detail 1

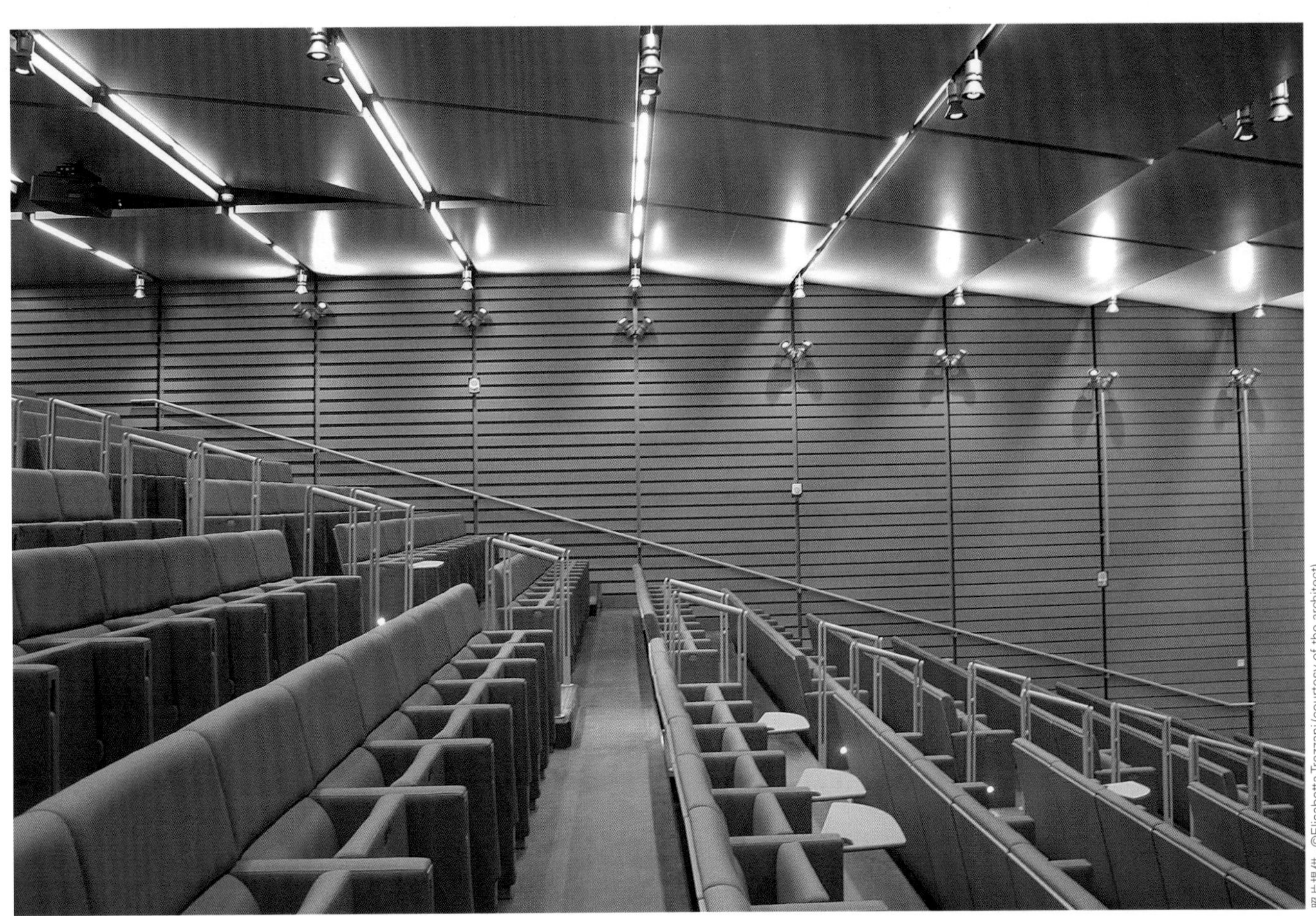

照片提供：©Elisabetta Trezzani (courtesy of the architect)

Comedie De Bethune——国家大剧院

Manuelle Gautrand Architecture

PALACE
LE PALACE

自20世纪80年代起，文化便成为北加莱海峡地区的发展中心。而La Comédie De Bethune项目则反映了利用国家剧院中心来表现该区域的代表性建筑的意愿。这里是一处具有参照意义且充满了艺术才能的地方，剧院全方位地展示其形式，并且亲近所有的观众，致力于打造吸引人们，再现该城市最具代表性的文化设施。

剧院由Vauban建造，位于华盖之下，设有300个坐席，并在1993年迁至一座古老的火药库之内。之后该剧院又进行了翻新，成为一座工作室剧院，并且沿用至今。那里曾有一座最初被称为是"宫殿"的影院，建于20世纪30年代，位于La Comedie De Bethune的现有场地内。其梁柱支撑的建筑立面全部移除，并于20世纪90年代拆毁。今天，街角的住宅也已拆除，而在16个月的施工工期结束后，新扩建的结构便呈现在人们的眼前，并且占据了部分的场地。

扩建

扩建的结构位于场地的一角，为三层建筑：这是一座多层建筑，与天然地基位于相同的水平面，一层设有彩排室，二层为办公层，连接着现有的办公层和原建筑的二层。彩排室的规模能够允许大厅搬至两条路相交的拐角处，这是1994年竞标设计中便存在的条款。该位置有几个优势：望向剧院的视野更佳，且通往咖啡吧的道路更便捷。咖啡吧之前位于彩排室之下，人们既无法从Victor Hugo大街直接进入其中，同时其本身也不能自然采光。而现在，位于角落的大厅与咖啡吧直接相连，在带来自然光照的同时，也形成了一条充满生气和欢迎氛围的通道。由于对现有剧院的二层进行了扩建，顶层的行政区便增添了一处大型且功能性更强的空间。

重建现有建筑

自15年前建筑交付以来，安全条例不断地进行了细致的规范，因此在现有的建筑内，很多空间都发生了变化。该项目因此也需要对整座建筑进行升级。一些安全条例还涉及到了消防条例和剧院内所有空间的残疾人通道。扩建的结构要求楼梯位于一层，将来建造的大厅的右侧，来为其上层的国家戏剧艺术学院（CNAD）的彩排室、将来其他的扩建结构以及二层的办公室服务。最后，该建筑还达到了最重要的保温和能效条例的要求。

正厅后排上面的房间

该事务所全身心地来完成这个房间的建造，且已脱离其原始的特色（在对原有建筑进行施工的过程中），它将被改造为另一间彩排室，没有网格，比主彩排室的规模小，但是人们能够通过办公层的楼梯向上到达。

立面

立面的概念须融入原始项目（带有平滑转角的大型圆形体量，漆成紫色）的高度可见的部分中。最后，立面呈现为黑色，一种深色且富有表现力的颜色。为了与原始项目之间建立一种软连接，黑色以波状金属面板的形式展现出来。而这些波状面板为大型菱形图案，使人想起了带有紫色圆形图案的面板。

现有的建筑如同组装在一个斜框架中，以安装这些特定形状的面板。这些波状面板或暗淡，或有光泽，且根据光照程度以及光线角度，来接收不同的日光量。

Comedie de Bethune - National Drama Theater

Since the 1980s, culture has become a priority in the development of the Nord-Pas-de-Calais area; "La Comedie De Bethune" reflected the will of the elected representatives of the department

东北立面 north-east elevation

to be equipped with a national drama center. As the place of reference and artistic excellence where the theater can express itself in all its forms and reach all audiences, the building contributes strongly to the attractiveness of the area and represents the most emblematic cultural equipment of the city. Installed in a marquee with three hundred seats in 1992, the theater moved in 1993 in an old powder magazine, built by Vauban and refurbished into a "Studio Theater" which is still used nowadays. There was initially a movie theater called "The Palace" built in the 1930s on the present site of "La Comedie De Bethune". The facade of the building maintained by girders, was left when it was demolished in the 1990s. Today, the street corner house having been demolished, the extension was able to be realized after 16 months of work, partially in occupied site.

The Extension

The extension is located on the corner of the plot, on three levels: a volume of high-rise located on the same level as the natural ground, and dedicated to the rehearsal room; then a level of offices, which links with the existing office floor to the second floor of the building. The size of the rehearsal room allows relocating the hall on the corner of the two roads, a provision that already existed during the competition in 1994. This position has several advantages: it allows to give a better visibility to the theater and to make the coffee bar more accessible. The coffee bar formerly located under the volume of the room's floor, could not benefit neither of a direct entrance from the Avenue Victor Hugo, nor a natural lighting. Now, the construction of the hall corner connects directly to the coffee, bringing a natural lighting and making it an animated and welcoming passageway. Above, the administration has a larger and more functional space thanks to the extending from the second floor of the existing theater.

Restructuring of the Existing

Many spaces have also been changed within the existing building. Indeed, since the delivery of the original building 15 years ago, the safety regulations have evolved considerably. This project therefore involved the upgrading of the entire building. Several safety regulations were concerned, including the fire safety regulations and the accessibility to all spaces of the theater for disabled people. The extension also required the construction of a staircase located on the right of the future hall and serving on the first floor the upper level of the rehearsal room of the CNAD (National Academy of the Dramatic Arts) and its additional premises, and the offices on the second floor. Finally, the building reached the most

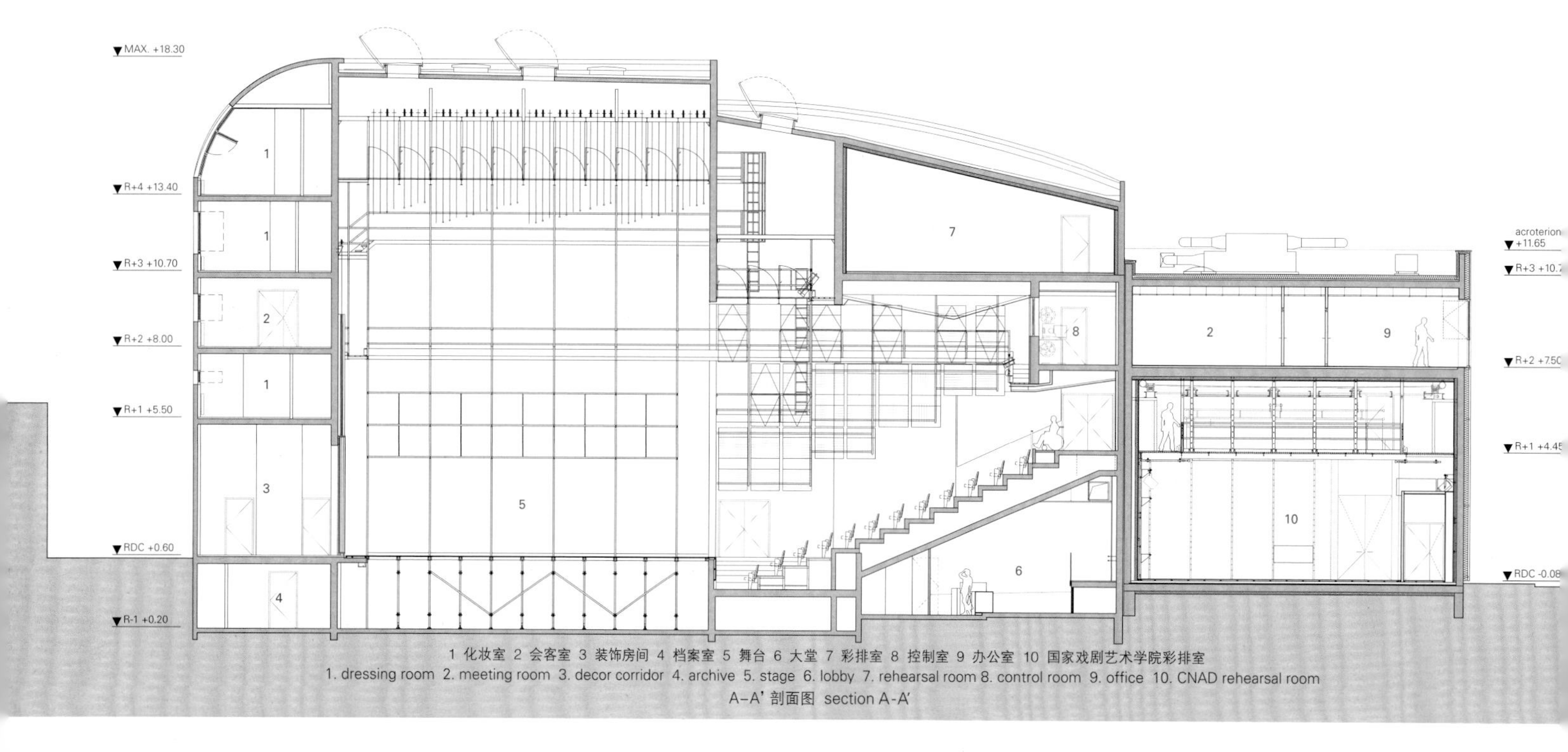

1 化妆室 2 会客室 3 装饰房间 4 档案室 5 舞台 6 大堂 7 彩排室 8 控制室 9 办公室 10 国家戏剧艺术学院彩排室
1. dressing room 2. meeting room 3. decor corridor 4. archive 5. stage 6. lobby 7. rehearsal room 8. control room 9. office 10. CNAD rehearsal room
A-A' 剖面图 section A-A'

项目名称：Bethune National Drama Theater
地点：138 Rue du 11 Novembre, 62400 Locon, France
建筑师：Manuelle Gautrand Architecture
结构工程师：Khephren / 舞台设计师：Bati-Scene
音效工程师：Jean-Paul Lamoureux / 流体工程师和经济学家：Hexa Ingenierie
行程安排与协调：HD Project / 控制：Preventec
承包授权机构：Communaute D'agglomeration de Bethune
使用者：La Comedie de Bethune / 表面积：total_2,082m², extension_720m²
扩建和改建造价：EUR 3.6 million
竞赛时间：initial building_1994, extension-restructuring_2009
施工时间：initial building_1999, extension-restructuring_2013~2014
摄影师：©Luc Boegly (courtesy of the architect) (except as noted)

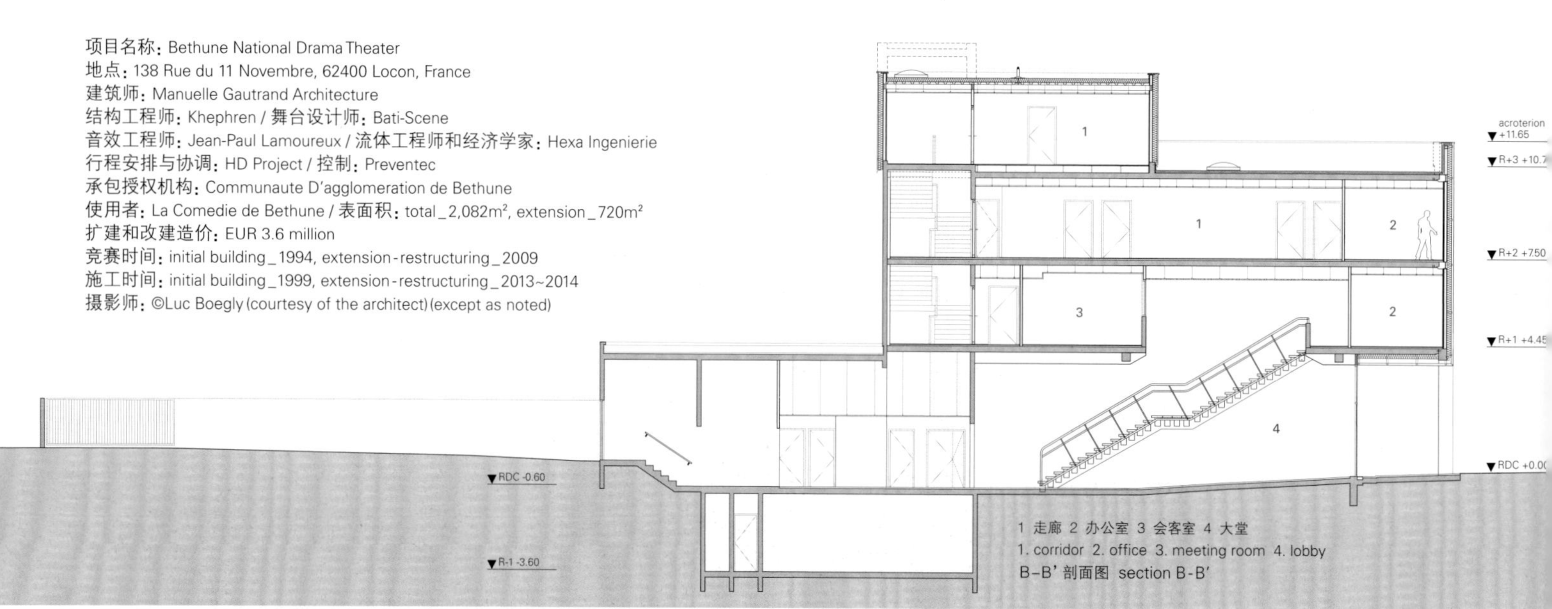

1 走廊 2 办公室 3 会客室 4 大堂
1. corridor 2. office 3. meeting room 4. lobby
B-B' 剖面图 section B-B'

ambitious thermal and energy regulations.

The Room Above The Pit

The agency used all these work to complete this room, which had been left "raw" during the construction of the original building, into a second rehearsal room without grid, and smaller than the main rehearsal room, with an access facilitated by the raising of the staircase of the offices.

The facades

The conception of the facades had to integrate the highly visible part of the initial project, this large rounded volume with the smooth corners, fully lacquered in purple. Finally, it's in black that the agency coated this extension: a deep and powerful color. To create a soft link with the initial volume, this black is implemented in the form of a kind of weaving metal panels, which are drawing large rhombuses, to remind the ones of the purple rounded shape.

The existing building is assembled on a diagonal frame which allows attaching those shaped panels. The waves of these panels, mat and glossy, take the light differently, depending on their degree of shine and their orientation to the light.

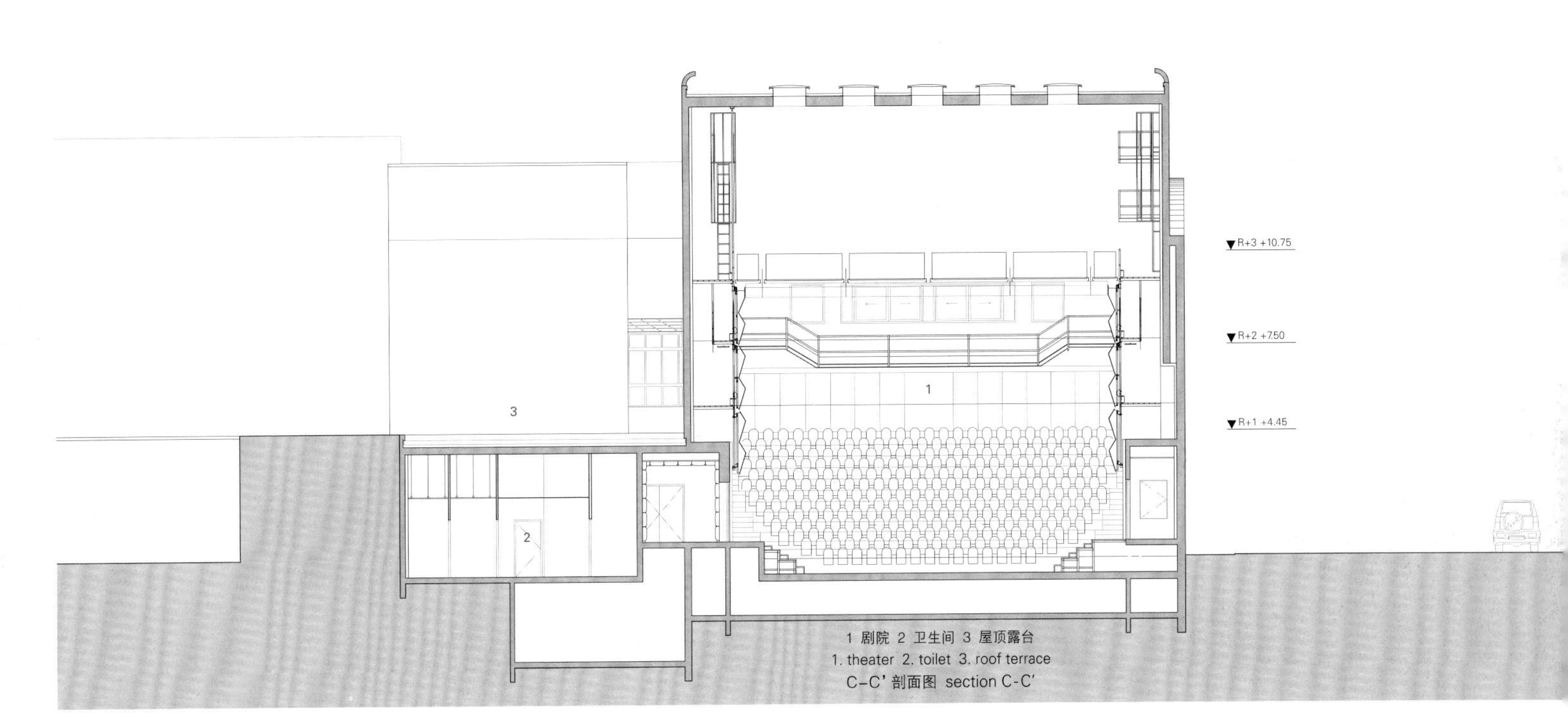

1 剧院 2 卫生间 3 屋顶露台
1. theater 2. toilet 3. roof terrace
C-C' 剖面图 section C-C'

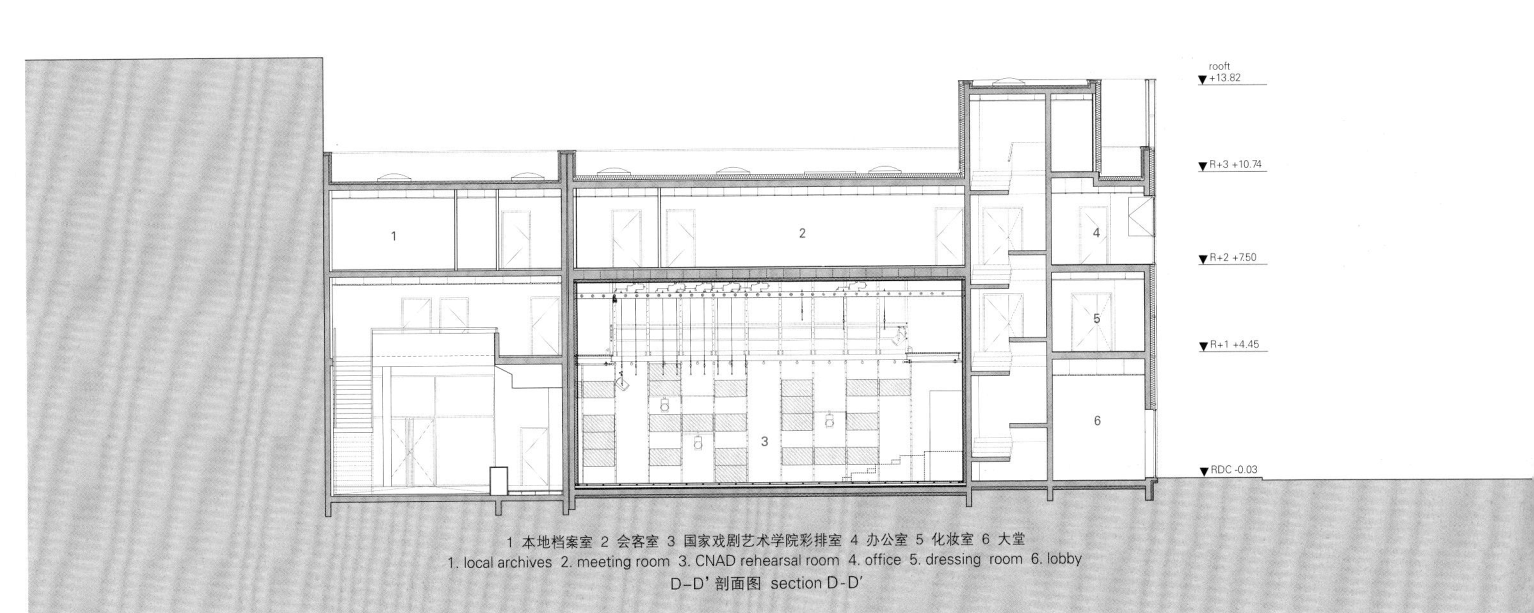

1 本地档案室 2 会客室 3 国家戏剧艺术学院彩排室 4 办公室 5 化妆室 6 大堂
1. local archives 2. meeting room 3. CNAD rehearsal room 4. office 5. dressing room 6. lobby
D-D' 剖面图 section D-D'

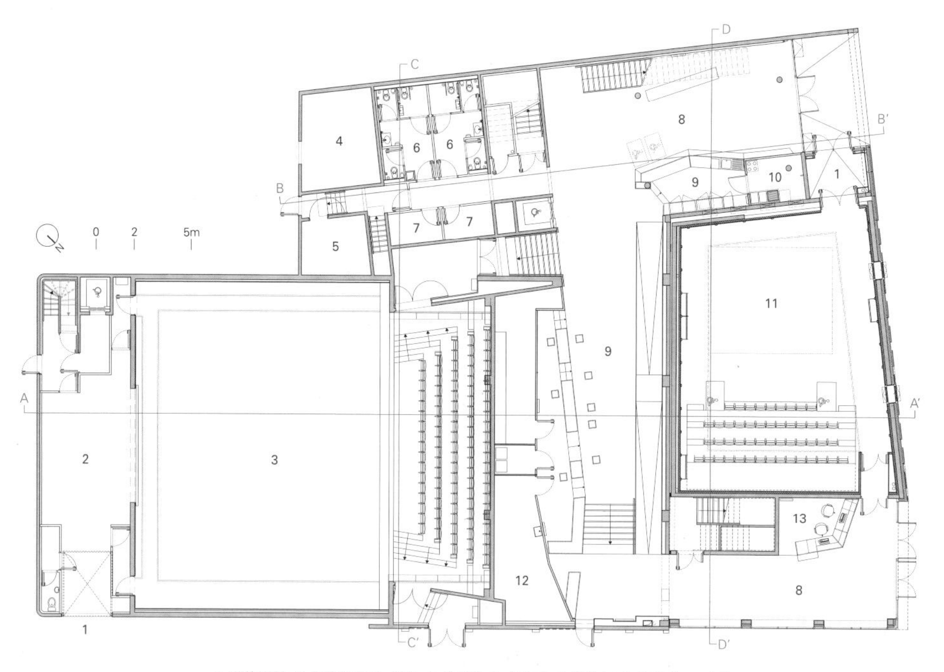

1 装饰通道 2 装饰走廊 3 舞台 4 变电室 5 电气室 6 卫生间 7 寄存处 8 大堂
9 自助餐厅 10 办公室 11 国家戏剧艺术学院彩排室 12 存储室 13 票务处
1. decor access 2. decor corridor 3. stage 4. transformer room 5. electricity room 6. toilet
7. clock room 8. lobby 9. cafeteria 10. office 11. CNAD rehearsal room 12. storage 13. ticket office
一层 first floor

原始剧院 previous theater

东北立面 north-east elevation

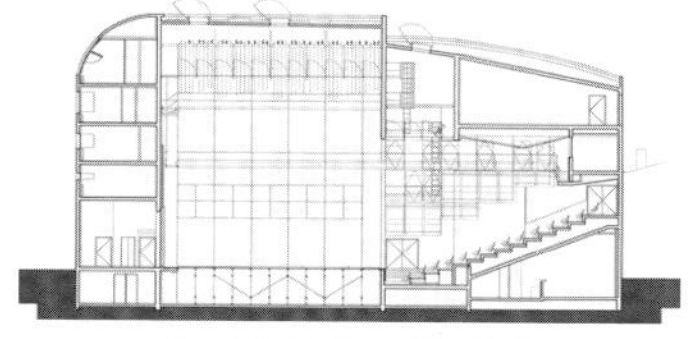

A–A' 剖面图 section A-A'

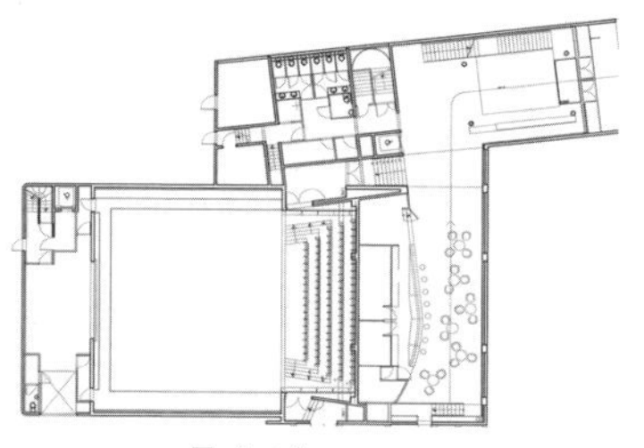

一层 first floor

照片提供：©Philippe Ruault (courtesy of the architect)

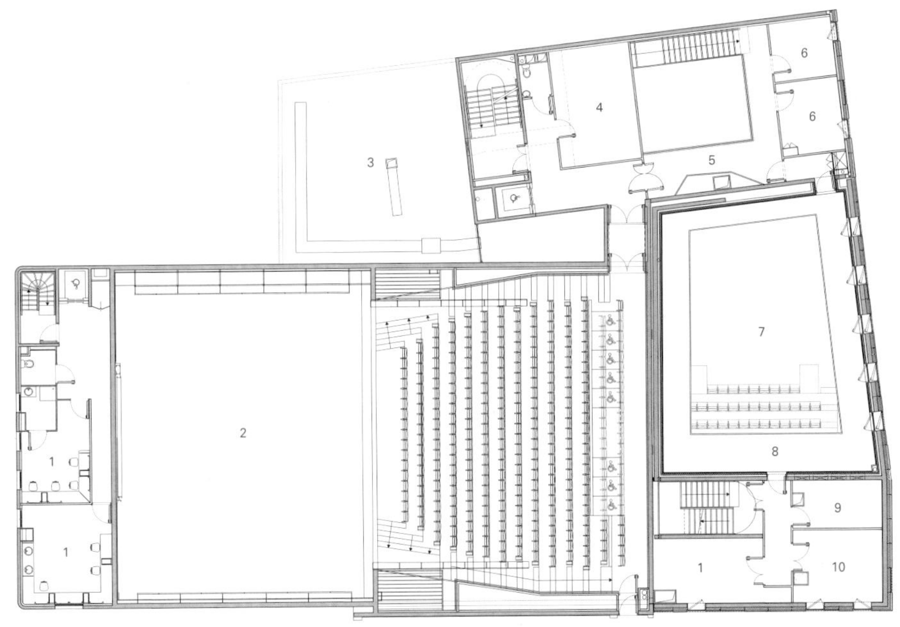

1 化妆室 2 舞台上空 3 屋顶露台 4 会客室 5 大堂 6 办公室
7 国家戏剧艺术学院彩排室 8 桥 9 电气室 10 存储室
1. dressing room 2. stage void 3. roof terrace 4. meeting room 5. lobby
6. office 7. CNAD rehearsal void 8. bridge 9. electricity room 10. storage
二层 second floor

>>102

Henning Larsen Architects

Is an international architecture company in Denmark, with strong Scandinavian roots. Founded by Henning Larsen in 1959, and is currently managed by CEO Mette Kynne Frandsen and Design Director Louis Becker. It has offices in Copenhagen, Oslo, Munich, Istanbul, Riyadh, the Faroe Islands and a newly established office in Hong Kong of China. Its goal is to create vibrant, sustainable buildings that reach beyond itself and become of durable value to the user and to the society and culture that they are built into. Its ideas are developed in close collaboration with the clients, users and partners in order to achieve long-lasting buildings and reduced life-cycle costs. This value-based approach is the key to our designs of numerous building projects around the world-from complex master plans to successful architectural landmarks. Won the prestigious European Union Prize for Contemporary Architecture-Mies van der Rohe Award 2013.

>>120

mlzd

Was set up in Bienne, Switzerland in 1997 as team of several architects. The work covers a very broad range, from urban development to matters of detail and from industrial projects through to museum ones. They have deliberately opted for variety in staffing their office of around twenty with individuals representing different cultures, ages and experience which promotes a multifaceted office culture. The lively internal debate gives birth to many different kinds of project, but these have one element in common: total dedication to the location and job entrusted to them. Has already won more than fifty competition awards, and more than twenty of these projects have also been realized.

>>172

Manuelle Gautrand Architecture

Is a Parisian based architecture firm founded in 1991 by Manuelle Gautrand who was born in France in 1961. She has received a lot of national and international Architecture Awards for several of her buildings and was awarded the Legion of Honor by the French Government in 2010. Her team of over 20 architects develops projects for public contracting authorities as well as private firms both in France and abroad.

>>36

Rafael Hevia left

Received an Architecture degree from the Pontifical Catholic University of Chile(PUC) in 2000. Has worked since then till 2007 at Mathias Klotz's office. Later he runs his own office, designing buildings, houses and furniture. Currently teaches at the Architectural Faculty of Diego Portales University(UDP).

Rodrigo Duque Motta right

Was born in 1975 in Santiago, Chile. Received an architecture degree from Pontifical Catholic University of Chile in 2001. Has worked in Mathias Klotz's office as project leader from 1999 to 2003 and he established his own office Rodrigo Duque Motta Arquitectos. Was awarded the Youths Class Awards of the Architecture's College as the best Chilean architect under 35 years old. Currently is a studio professor at the Diego Portales University.

>>50

Legorreta + Legorreta

Victor Legorreta was born in Mexico City, 1966 and studied Architecture at the Ibero-American University in Mexico City. After practicing in several architectural firms, he joined Legorreta Arquitectos in 1991 and the name of the firm changed to Legorreta + Legorreta in 2000, representing a new organization. Is now a partner and design leader at Legorreta + Legorreta. Has lectured in numerous universities and conferences in the USA. and Latin America. Has also been a member of several juries, such as Design competitions for the American Institute of Architects(AIA). In 2007, Victor is awarded as an Honorary Member of AIA.

>>18

CRAB Studio

Peter Cook[right] was born in Southend-on Sea in 1936. Studied at Art University Bournemouth and Architectural Association School of Architecture in London. Was knighted by the Queen in 2007. Is a Royal Academician and a Commander of the Order of Arts and Letters and a Senior Fellow of the Royal College of Art, London. Gavin Robotham[left] was born in North Walsham in 1969. Studied at Bartlett School of Architecture and Harvard Graduate School of Design. Is the director and design team leader of CRAB Studio and continues to lecture design and acts as a consultant.

>>10

Heatherwick Studio

Was founded by Thomas Heatherwick in 1994. Today his team works in King's Cross on projects ranging from bridges and buildings to products and large scale works of public art. Has won the Prince Philip Designers Prize in 2004. Has been awarded Honorary Doctorates from the Royal College of Art, University of Dundee, University of Brighton, Sheffield Hallam University and University of Manchester. Is an Honorary Fellow of the Royal Institute of British Architects and was elected a Royal Academician by the Royal Academy of Arts, London in 2013.

>>132

Flores & Prats

Is an architecture office from Barcelona, dedicated to confront theory and academic practice with design and construction activity. After their experience at Enric Miralles' office, Ricardo Flores and Eva Prats have developed a career where research has been always linked to the responsibility to make and to build. The office has worked on rehabilitation of old structures for new occupations, as well as on neighbors' participation in the design process of urban public spaces, and on social housing and its capacity to create community. Their work on rehabilitation Casal Balaguer Cultural Center was exhibited at the Venice Biennale of Architecture 2014 and has just been nominated for the Mies van der Rohe Award 2015. They are currently teaching at the School of Architecture of Barcelona.

Duch-Pizá

Was founded in 1990 by Sé Duch Navarro and Xisco Pizá Alabern. Forms a multi-disciplinary team that offers a fast and adapted service to the circumstances of each case. Has extensive experiences in public and private works, as well as in dealing with the government projects. Participates in the ground search, handles the complete processing of projects and offers comprehensive management of the execution of the works.

>>60

BUSarchitektur

Was founded in 1986 in Buenos Aires by Claudio J. Blazica(1956~2002) and Laura P. Spinadel. Laura was born in Buenos Aires, Argentina in 1958. Studied architecture and received Gold Medal at the University of Buenos Aires. In Vienna, the architectural firm has been run jointly with Jean Pierre Bolivar and Bernd Pflüger since 2003. BOA office for advanced randomness was established as a collaboration between Laura P. Spinadel and Hubert Marz in order to encourage alternative approaches to the communication and discussion of spaces and urbanities. Their works were nominated and honored at the numerous awards including Mies van der Rohe, ArchDaily Award, European Award AIT and Award BHP.

>>88

Richard Kirk Architect

The Principal Director, Richard Kirk is a Registered Architect of Queensland and APEC. He established Richart Kirk Architect in 1995 after graduating from the University of Queensland in 1991 with first class Honors. He received a Bachelor of Design Studies and Architecture there. His noted academic achievements provide an intellectual basis for the foundation and direction of the practice which now has studios located in Brisbane, Kuala Lumpur and Beijing. He is the current Australian Institute of Architects(AIA) Queensland Chapter President, Fellow of the AIA, Member on the Board of Architects of Queensland, Adjunct professor at University of Queensland, and Member of the Brisbane City Council Independent Design Advisory Panel.

HASSELL

Mark Loughnan is a head of design for Architecture and a Board Director at HASSELL, based in Melbourne. Mark joined HASSELL in 2006 and has a diverse history in a wide range of building typologies with a focus on public and civic architecture. He is currently involved in a large range of projects and management aspects across the practice. Prior to joining HASSELL, Mark spent seven years with Herzog & de Meuron, winners of the Pritzker Prize in 2001. Mark also established and managed Herzog & de Meuron in San Francisco, USA, becoming an associate of the firm in 2004.

Angelos Psilopoulos
Studied architecture at the School of Architecture, Aristotle University of Thessaloniki(AUTh), then moved on to his Post-Graduate studies at the National Technical University in Athens(NTUA). Is currently pursuing his Ph.D. at the NTUA on the subject of Theory of Architecture, studying gesture as a mechanism of meaning in architecture. Has been working as a freelance architect since 1998, undertaking a variety of projects both on his own and in collaboration with various firms and architectural practices in Greece. Since 2003, he has been teaching Interior Architecture and Design in the Department of Interior Design, Decoration, and Industrial Design at the Technological Educational Institute of Athens(TEI).

Alison Killing
Is an architect and urban designer based in Rotterdam, the Netherlands. Has written for several architecture and design magazines in the UK, contributing features and reviews to Blueprint and Icon and editing the research section of Middle East Art Design and Architecture. Most recently, she has worked as a correspondent for the online sustainability magazine Worldchanging. Has an eclectic design background, ranging from complex geometry and structural engineering, to humanitarian practice, to architecture and urban design and has worked internationally in the UK and the Netherlands, but also more widely in Europe, Switzerland, China and Russia.

>>160

Renzo Piano Building Workshop
While studying at Politecnico of Milan University, Renzo Piano worked in the office of Franco Albini. After graduating in 1964, he started experimenting with light, mobile,temporary structures. Between 1965, and 1970, he went on a number of trips to discover Great Britain and the United States. In 1971, he set up the Piano & Rogers office in London together with Richard Rogers. From the early 1970s to the 1990s, he worked with the engineer Peter Rice. Renzo Piano Building Workshop was established with 150 staff in Paris, Genoa, and New York.

©Edward Tyler

>>76

Stanton Williams Architects
Was founded by Alan Stanton[second] and Paul Williams[fourth] in 1985. Paul Williams has been a visiting critic and lectured at several universities and institutes in the UK and abroad. He is currently an examiner at Birmingham University School of Architecture. Patric Richard[fifth] is passionate about the relationship between landscape and architecture, the urban realm and the role of architecture as a catalyst for regeneration. Alan Stanton has been a vice president of the architectural association council. He has lectured extensively in this country and abroad and currently he is an external examiner at Greenwich University. He also believes that a great architecture is a profound understanding of space, and that context and heritage are creative forces in design. Peter Murray[first] studied architecture at Melbourne University and the Polytechnic of Central London. Gavin Hendersonth[third] has been a visiting critic and lecturer at a number of UK schools of architecture. Has a particular interest in the development and culture of cities and the role of public space.

>>146

noAarchitecten
Was established in 1999 by An Fonteyne[middle], Jitse van den Berg[right] and Philippe Viérin[left]. They currently teach in Belgium and abroad including TU Delft and are invited as guest critics at international schools and institutions. An Fonteyne(1971, Ostend) graduated from Ghent University in 1994 and worked for DKV Architects and David Chipperfield Architects. Jitse van den Berg(1971, Nijmegen, NL) studied at TU Berlin and TU Delft. He worked for Sauerbruch Hutton Architects and David Chipperfield Architects. Philippe Viérin(1969, Bruges) graduated from Ghent University in 1992 and worked for Stéphane Beel and KCAP. He also runs Viérin Architecten in Bruges.

著作权合同登记06-2015年第59号

图书在版编目(CIP)数据

地域文脉与大学建筑 : 汉英对照 / 韩国C3出版公社编 ; 史虹涛等译. —大连: 大连理工大学出版社, 2015.6
(C3建筑立场系列丛书)
书名原文: C3 University Buildings in Context
ISBN 978-7-5611-9885-8

Ⅰ. ①地… Ⅱ. ①韩… ②史… Ⅲ. ①高等学校—建筑设计—汉、英 Ⅳ. ①TU244.3

中国版本图书馆CIP数据核字(2015)第117141号

出版发行：大连理工大学出版社
(地址：大连市软件园路 80 号　邮编：116023)
印　　刷：上海锦良印刷厂
幅面尺寸：225mm×300mm
印　　张：11.75
出版时间：2015 年 6 月第 1 版
印刷时间：2015 年 6 月第 1 次印刷
出 版 人：金英伟
统　　筹：房　磊
责任编辑：许建宁
封面设计：王志峰
责任校对：高　文

书　　号：978-7-5611-9885-8
定　　价：228.00 元

发　行：0411-84708842
传　真：0411-84701466
E-mail：12282980@qq.com
URL: http://www.dutp.cn